P9-CJH-937

OTHER TITLES OF INTEREST FROM ST. LUCIE PRESS

Development, Environment and Global Dysfunction

Toward Sustainable Recovery

Yosef Gotlieb
Research Faculty
The George Perkins Marsh Institute
Clark University
Worcester, Massachusetts and
Director
The Israel Center for International
Environmental Studies (ICIES)

St. Lucie Press
Delray Beach, Florida

Printed and bound in the U.S.A. Printed on acid-free paper.
10 9 8 7 6 5 4 3 2 1

ISBN 1-57444-012-8

Phone: (407) 274-9906
Fax: (407) 274-9927

StL

Published by
St. Lucie Press
100 E. Linton Blvd., Suite 403B
Delray Beach, FL 33483

TABLE OF CONTENTS

PREFACE .... **ix**

CHAPTER ONE—INTRODUCTION **1**
Terms of Discourse 4
Progress and Growth 9
The Consumption of Nature 10
Reconceptualizing Development 11
The Development of Global Dysfunction 13
Social Ecology against Development 20
The Organization of this Volume 21
Notes 22

CHAPTER TWO—GLOBAL DYSFUNCTION: MULTIPLE PROBLEMS, ONE PROBLEMATIC **25**
Decline in Third World Development 26
Environmentally Unsustainable Development 28
The Endurance of Ethnic Conflict 32
Socio-Spatial Realities of Territory 34
Problems of State 35
Modernization 36
The Limits to State-Centered Modernization 38
"Development" and Global Dysfunctional 39
Notes 42

CHAPTER THREE—THE SOCIETY-NATURE RELATIONSHIP **43**
Extremes Defining the Spectrum of Symbiotic Human-Land Relations 44

The Spectrum of Symbiotic Approaches to Society-Environment Relations 45
Human Ecology 45
Culture and Nature 45
Ecological Anthropology 47
Cultural Ecology 49
The Ecosystem Approach 49
Regional Political Ecology and Political Ecology 51
Bookchin's Social Ecology 53
Ethnoscience, Indigenous Knowledge, and Eco-Development 55
Intimacies of Society-Land Unity 56
Ethnoscience and Ethnoecology 59
Ethnoscience and Gender 60
The Integrity of Life-Place 62
Notes 64

CHAPTER FOUR—THE ENDOGENOUS RECOVERY REGION 69
The Rurality of Development 72
Bottom-Up Regional Development 73
Territorial Development 74
Agropolitan Development 75
The Politics of Place 77
Ecodevelopment 79
The Recovery of Territorial Life 80
Spatial Parameters of the Endogenous Recovery Region 83
Questions of Scale 84
Notes 85

CHAPTER FIVE—ARTICULATION OF THE SOCIAL ECOLOGY THESIS 87
The Material Manifestations of Ethno-Nationalism 88
Lessons of The Society-Environment Relationship 91
Colonized Space and the Impediments to Development 92
The Social Ecology Program 98

Social Ecology as Political Struggle 104
Notes 105

CHAPTER SIX—CONTEMPORARY KURDISTAN AS AN ENDOGENOUS RECOVERY REGION 107
Dimensions of Kurdish Identity 107
Demography of the Kurds and of Kurdistan 108
Modernization, the Turkish State and the Kurds 112
De-Kurdification and the Southeast Anatolia Project 116
Kurdistan in Iraq: Internal Colonialism 118
Oil and the Iraqi State Center 120
Iranian Kurdistan in Numbers 121
Kurdistan as an Ecoregion 124
Notes 129

CHAPTER SEVEN—ENDOGENOUS RECOVERY AND A CHOICE OF FUTURE(S) 131
Social Crises 132
The Loss of Wisdom, the Loss of Self, the Loss of Community 133
A Cosmopolitan World 134
Environmental Degradation 135
What We Do Know 136
History Reconstructed 138
Technology and Liberation from Nature 140
Recovering Reality 141
Recovery: Multiple Trajectories 143
Recovering Place 144
A Fuller View of The Parts of The Whole 145
Notes 146

BIBLIOGRAPHY 147

INDEX 183

For the Deliverers of my Past,
my Mother and Father

For my Life-Partner, Dr. Frida Grynspan,
who makes the passage of this life a Blessing

For my children, Sara Avital ("Tali") and Amihai,
the joys of the present who are my future

PREFACE

In the winter of 1987 I served as part of a team of Nepalese and Israeli planners assigned to formulate an integrated rural regional development plan for the Dang district in the Inner Terrai. The planning project followed four months of lectures, clinics and touring by an international group of African, Asian and Caribbean development professionals attending a course in Israel on integrated regional planning. I had directed the course for several years and was pleased by the opportunity to accompany the students and participate along with my colleagues on the multinational, interdisciplinary planning team.

Prior to my departure for Nepal, I considered myself relatively well-prepared for the project. I had seen poverty before in Costa Rica where I was born, in Chicago's South Side where I spent my childhood, in Bedford-Styversant and the South Bronx, and in Jewish development towns and Arab villages in Israel. I was conversant with the development literature, and I felt that I had sufficient global perspective on the contemporary Third World in large measure due to interaction with colleagues from countries as varied as Belize, Cameroon, the Philippines and Sri Lanka. I had become conversant with the multiple indicators and statistics used by the World Bank and other bodies to describe the stubbornness of the development struggle. I was ready to collaborate in that struggle.

Neither the course participants, my counterparts, nor I were prepared for the poverty we encountered in Nepal. Destitution in that society included but also exceeded material want. It was an impoverishment so profound that those in its chokehold were unable to conceive of options for a better life. The people I saw were disinherited and dispirited, caught in the wheels of Modernization without the means to contend with it. They were part of the one billion people on this planet who comprise the "absolute " poor.

Not only did the people of Nepal suffer, but their environment seemed deeply wounded as well. The inextricability of development and ecology was hammered home during my assignment. The weather had warmed considerably, and everything seemed awash in brilliant sunlight. I joined some of the course participants on the roof of our guesthouse, which had a magnificent view of the mountains surrounding Kathmandu. The mountains were majestic, but also disrobed. The treeline was at the summit of the range, and one could clearly count the stands of trees—in places even individual tress could be discerned. While commercial logging was going on elsewhere in the country, the forests of the Kathmandu Valley had disappeared under the weight of an increasing population. In one instant the fuelwood crisis described in the professional literature was made starkly graphic. More people meant more mouths to feed and more bodies to keep warm. Since fuel for cooking and heating came mainly from wood burning (the rest was generated from dried cow dung), it was inevitable that the forests would be harvested. The harvest had run out of trees in many places and unfurled over an ever-widening expanse. Similarly, Nepal's abundant waters had been injured by human and animal wastes, silting due to shifting soils, and the re-routing of waterways for irrigation. The replenishment cycle was destabilized at the "roof of the world" where melting Himalayan snowcaps formed the head waters for the Ganges and numerous other rivers on which the entire Indian sub-continent depends.

My experience in Nepal and subsequent ones challenged many of the assumptions implicit to the development enterprise. Many of these assumptions and refutations are value judgements. But if we accept one premise alone—that future generations are entitled to a viable biosphere—evidence strongly suggests that global change has been dysfunctional and unsustainable—socially and environmentally.

This books represents my attempt to understand the pathos we experience in a world that is as technologically advanced and creative as our own. Analytic clarity comes only when we "de-learn" convention and look at these seemingly distinct sets of phenomena as being inter-connected and mutually formative.

My hope is that this book will prove instructive to those committed to progressive socioeconomic change and a sustainable biosphere. My message contained here has been made clearer by Sandra Koskoff, the editor who shepherded the manuscript to St. Lucie Press less than a year ago and who has contributed to every stage of production. She has my gratitude as does my publisher, Dennis Buda, who welcomed the project on behalf of St. Lucie.

My thanks are extended to teachers, students, colleagues and friends too numerous to mention individually.

Special appreciation is expressed here for the people of Nepal and Kurdistan—and all other ancient peoples who, like my own, are determined to survive.

About the Author

Dr. Yosef Gotlieb is the founding Director of the Israel Center for International Environmental Studies (ICIES), a non-governmental organization aimed at representing Israel in the international movement for environmental sustainability.

Currently a member of the research faculty at Clark University's Marsh Institute, Dr. Gotlieb has taught courses on development and environmental studies at Clark, Rensselaer Polytechnic Institute and the Development Study Center (DSC), Rehovot, Israel, where he directed courses on development for students from Africa, Asia, and the Caribbean. Yosef Gotlieb was born in Costa Rica and has conducted field research in Nepal and Israel. He has served as co-principal investigator on a project funded by the MacArthur Foundation on sustainable resource use in Western Siberia with colleagues from the Russian Academy of Sciences and from the Marsh Institute. His current research interests also include ecological energetics, indigenous knowledge systems, and environmental cooperation as a means to reduce international conflicts, particularly in the Middle East.

Among Dr. Gotlieb's publications is *Self-Determination in the Middle East* (Praeger 1982) and several other books and monographs. His articles have appeared in academic and popular journals. His volume *Dysfuncíon Muñdial* will be released in Spanish by the Center for International Political Economy and Sustainable development of the National University of Costa Rica.

CHAPTER 1

INTRODUCTION

Socioeconomic development has been a major preoccupation of governments, aid organizations and change agents since World War II. Although improvement has been recorded for *some* indicators in *some* societies *sometimes*, sustained development has been evasive. This failure has frustrated many development specialists who now indict development itself as contributing to deepening underdevelopment (Amin 1990; Barnett 1989; Friedmann 1992; Johnston 1989; Redclift 1987; Sachs et al. 1992; V. Shiva 1989, 1991; N. Smith 1990).

Development is criticized for a broad range of maladies including environmental degradation (Commoner 1990; Ehrlich and Ehrlich 1991; Leonard et al. 1989; W. Sachs 1992; WCED 1987; World Bank 1992; WRI, UNDP and UNEP 1992), the feminization of poverty and the patriarchal tyranny in rural areas of developing societies (Kelkar and Nathan 1991; Shiva 1989), ethnocentrism and cultural destruction (Max-Neef 1991; T. Verhelst 1990), disempowerment and exploitation resulting from rural-urban outmigration (Friedmann 1992; Gugler et al. 1988) and ethno-political violence surging across the borders of post-colonial states (Cobbah 1988;

Gotlieb 1992a, b; Hirschman 1981; Hughes 1991; Seers 1983; Tinker 1981). On these grounds and others many veterans of the development struggle would concur with W. Sachs that "The last 40 years can be called the age of development. This epoch is coming to an end. The time is ripe to write its obituary" (1992: 1).

Why has development failed? In reviewing the history of development as a deliberate enterprise,[1] it is apparent that policies and programs instituted are highly diverse. Development theory has undergone constant revision (Figure 1.1) and has been informed by a broad spectrum of approaches. One can only conclude that the problems of development are larger than the highly heterogeneous policies, plans and programs undertaken by the full spectrum of economic approaches. It is increasingly evident that our understanding of the reasons why one fifth of humanity enjoys unprecedented wealth while the other four-fifths live in various states of poverty and privation is at best incomplete. Clearly, conventional analyses of socioeconomic, cultural, ecological and political phenomena are limited in their ability to elucidate the multiple problems we now encounter globally.

The thesis presented here is that development itself, or at least the concepts we use to define it, are deficient. We can speak of a global problematic embodying three broad sets of phenomena: *trenchant poverty, environmental degradation* and *socio-political unrest*. Related to this problematic is existential malaise, particularly in those societies considered the most "developed" (Roszak 1992; Young 1990). This global problematic is not transient. It speaks to profound issues about who we are that have become repressed in our individual and social consciousness.

I contend that the development problematic derives from an approach to modernity—*Modernization*[2]—that is socially and ecologically maladaptive. Modernization is defined here as the interwoven processes of urbanization, industrialization and secularization that occurred following the Industrial Revolution. The concept is expanded in the following chapter. Insofar as development is defined by the Modernization paradigm, it is an enterprise working *against nature*.

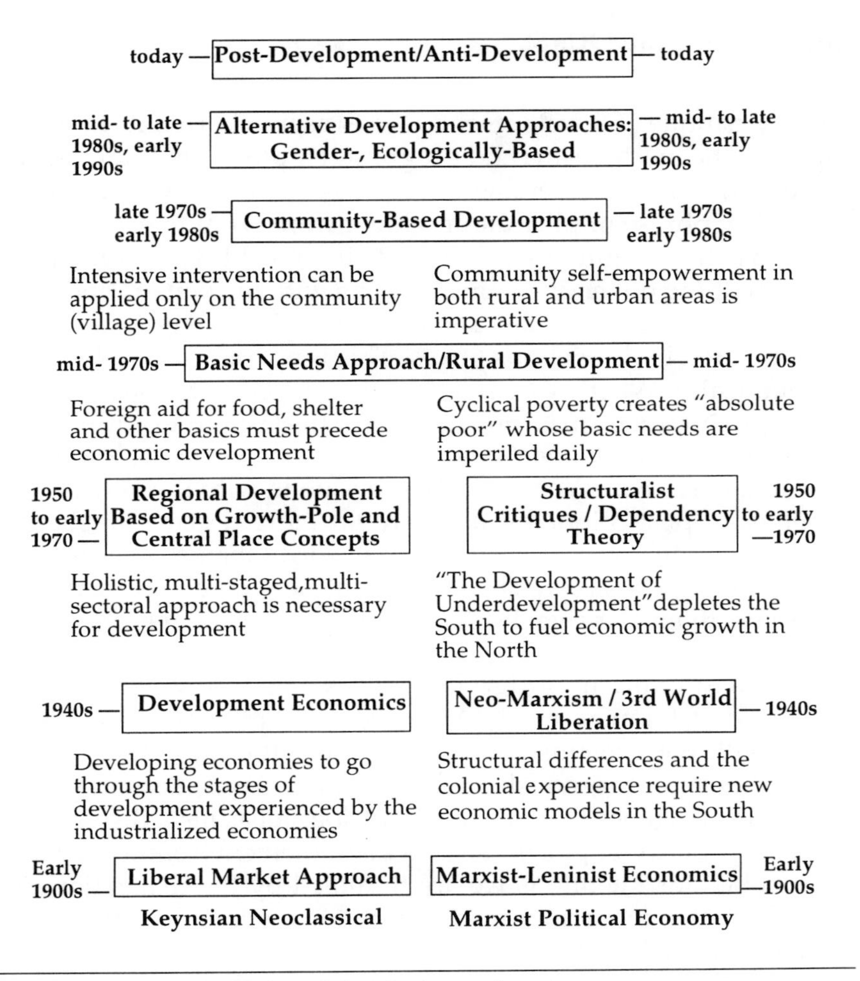

Figure 1.1 The evolution of development theory.

Inherent in its construction of "rational" space economies, development qua Modernization replaces the unique relationship between social groups and their environments with a global pop-culture based on the production of needs and ever-expanding consumption.

TERMS OF DISCOURSE

My critique of development derives as much from its first principles as the mechanisms and programs employed to implement it. Among the most vulnerable concepts is the term "development" itself. The phrase implies an unfolding from within yet recent history has shown quite plainly that Third World development has been anything but endogenous. Starting with the penetration of the New World, African, Asian, and Latin American peoples have been manipulated by exogenous forces. This is as true culturally as it is economically. Whether these societies enjoyed autonomy prior to the rise of imperialism is immaterial. For the last three centuries and especially during the post-colonial period, power centers in Bonn, London, Paris, New York, and Tokyo—along with the new domestic elites of post-colonial states—are far more decisive in shaping the lives of the impoverished global majority than any indigenous factor among the latter. The creation of new core-periphery relations effectively deprive the disinherited of control over their lives.

My first criticism of development is based on its pretensions. Neither development nor its Modernization mission has anything to do with endogenous change; they are little more than ad hoc adaptations to exogenously imposed conditions. A second pretension is an arrogance that development specialists have displayed in their attempts to "engineer" societies and their environments. However noble our intentions we are humbled by the realization that our efforts have possibly made conditions worse than they were under the brutality of colonialism. In seeking to "develop" societies in the image of our own, we seem unaware that with respect to fundamental, integrated social processes, the West has as much to learn from ancient peoples, perhaps more than what we can offer them. A lack of such wisdom and a contempt for "primitive" practices and beliefs of rural peoples have led us to a kind of spiritual grave-robbing. Modernization has crushed the only wealth these people own: a heritage of community and continuity (Weitz 1986). Rejection of Modernization does not demand a rejection of all things modern, just as the call for social change (e.g., ethnic autonomy, human rights, gender parity) should not require peoples and individuals to abandon their traditional, historical identities.

Development is further damned by the terms of reference it uses. These terms reflect what M.A. Max-Neef has called the "impoverishment of . . . the development language" (1991: 99). As expressed by Sachs et al., "At a time when development has evidently failed as a socioeconomic endeavor, it has become of paramount importance to liberate ourselves from its domination over our minds" (1992: 5).[3] At the very least the word "development" has been so variously interpreted that it means little. Even the widely used phrase of "sustainable development," while representing an evolution in our consciousness of development and environment, is so broadly applied as to be rendered ludicrous. Surely the notion of sustainable development accepted by former American Vice-President Danforth Quayle at the 1992 Earth Summit is profoundly different and at odds with that of Zimbabwe's President Robert Mugabe or of former Costa Rican President Oscar Arias Sanchez.

Further representative of the problematic of development discourse is its categorization of societies (i.e., the social formations "captured by" state borders),[4] in terms of polarities—rich and poor, more developed and less developed, developed and developing, developed and underdeveloped, advanced market economies and newly industrialized countries, North/South, and the core/semi-periphery/periphery typology used in world-system theory (Taylor 1986; Wallerstein 1982). Given the absence of agreement on what constitutes development,[5] it is not surprising that typological terms used by the major international agencies to describe societies are frequently revised and differ from agency to agency (Table 1.1).

Implicit to the classification of relative development standing of societies are several assumptions:

- Development reflects the "natural" direction of human social change—the success of human agency is measured by increased capacity to "control" nature;

Table 1.1 Inconsistencies in Typologies of Development Status

World Development Report 1988	International Monetary Fund	United Nations	United Nations Conference on Trade and Development	General Agreement on Tariffs and Trade
Industrial market economies	*Industrial countries*	*Developed market economies*	*Developed market economies*	*Developed countries*
OECD (excludiing Greece, Protugal, and Turkey)	North America Canada USA	North America Canada USA	North America Canada USA	North America Canada USA
	Europe EC (excluding Greece and Portugal) EFTA	Europe EC EFTA	Europe EC EFTA	Europe EC EFTA
		Other Europe Faeroe Islands Gilbraltar Malta	Other Europe Faeroe Islands Gilbraltar	Other Western Europe
		Africa South Africa	Africa South Africa	Africa South Africa
	Asia Japan	Asia Israel Japan	Asia Israel Japan	Asia Australia Japan New Zealand
	Oceania Australia New Zealand	Oceania Australia New Zealand	Oceania Australia New Zealand	
Developing economies	*Developing countries*	*Developing market economies*	*Developing market economies*	*Developing economies*
Latin America and the Caribbean	Western Hemisphere	Americas (excluding North America)	Americas CACM CARICOM LAIA Other	Latin America
Europe (including Cyprus, Greece, Hungary, Malta, Poland, Portugal, Romania, Turkey, and Yugoslavia)	Europe		Europe Malta Yugoslavia	
Middle East and North Africa	Middle East (including Egypt)			Middle East

Table 1.1 Inconsistencies in Typologies of Development Status **cont.**

World Development Report 1988	International Monetary Fund	United Nations	United Nations Conference on Trade and Development	General Agreement on Tariffs and Trade
Developing economies	*Developing countries*	*Developing market economies*	*Developing market economies*	*Developing economies*
Sub-Saharan Africa	Africa (including South Africa	Africa North Other CEUCA ECOWAS Rest of Africa((excluding South Africa)	Africa North Other CEPGL CEUCA ECOWAS Other (excluding South Africa)	Africa (including South Africa
South Asia East Asia	Asia (excluding Middle East but including Oceania)	Asia Western Asia Other Asia Oceania	Asia Western Asia Other Asia Oceania	Asia (excluding Australia, Japan, New Zealand, and China and other Asian centrally planned economies)
High-Income oil exporters	*Twelve major oil exporters*	*OPEC*	*Major Petroleum exporters*	
Nonreporting nonmembers	*U.S.S.R. and other nonmembers not included elsewhere*	*Centrally planned economies*	*Socialist countries*	*Eastern trading area*
		Asia (including China) Europe and U.S.S.R. (including Hungary, Poland, and Romania)	Asia Eastern Europe (including Hungary, Poland, and Romania)	China and other Asian centrally planned economies Eastern Europe and U.S.S.R. (including Hungary, Poland, and Romania)
Other analytical groups				
Developing economies	*Developing countries*	*Developing countries*	*Developing countries*	*Eastern trading area*
Low-income China and India Other low-income	Low-income countries, excluding China and India	Least developed countries	Least developed countries	Least developed countries

Table 1.1 Inconsistencies in Typologies of Development Status **cont.**

World Development Report 1988	International Monetary Fund	United Nations	United Nations Conference on Trade and Development	General Agreement on Tariffs and Trade
Developing economies	*Developing countries*	*Developing countries*	*Developing countries*	*Eastern trading area*
Middle-income Lower middle-income Upper middle-income			Income groups based on 1980 GDP per capita: less than $500 $500–$1500 Over $1500	
Oil exporters Exporters of manufactured goods Highly indebted countries Sub-Saharan Africa	Oil exporters Exporters of manufactured goods 15 highly indebted countries Sub-Saharan Africa		Major exporters of manufactured products	15 highly indebted countries

Source: World Bank. 1994. *World Development Report 1989.* Oxford and NY: Oxford UniversityPress, 250–251.

- There is a specific trajectory to development—society can be ranked along this trajectory with respect to a uniform standard (e.g., economic growth, measures of social welfare);
- Anthropocentrism and the notion that human beings can exist apart from nature leads inevitably to the utilitarian and market-oriented value system that society places on its natural resources, landscape, life-place,[6] culture and collectivity;
- Development is achieved through increasing industrialization, urbanization and secularization, i.e., Modernization; and
- Successful development is growth-oriented and state-centered.

While development is widely assumed to result in socioeconomic advance, albeit it Eurocentrically defined (Barnett 1989), I believe that apparent progress has been more illusory than real, and more deleterious than beneficial to the survival of humanity and the ecosphere.[7]

PROGRESS AND GROWTH

The idea of progress is associated with greater freedom for the individual—freedom from material want, freedom to live one's life as one pleases unencumbered by the vestiges of social "primitivism" (traditions, community, etc.) of the past. The prevailing view is that "bettering oneself" through the acquisition of more goods and services is a blessing unavailable to earlier generations. The good of the individual is equated with the economic growth of society. Effectively, the objective of development is to meet needs through the increase of accumulated goods. Development entails the expansion of economic activity aimed at meeting needs and increasingly with providing amenities.

Development is, essentially, growth. The teleology of crude or enlightened capitalism—the preeminent force in the industrial global economy—is the promise of cornucopia in the foggy, distant future; the supposition of boundless plenty is implicit. But as we witness today, the sky literally is a limit to economic growth. So is the planetary supply of water, soil, flora, fauna and other stock and flow resources that are indispensable parts of our "life-support system" (Ehrlich and Ehrlich 1991). In a fundamental way contrary to conventional views, *growth-based development produces scarcity by consuming irrecoverable stock resources and by the degradation inherent to resource extraction, processing and production.*[8]

As noted by Ehrlich and Ehrlich (ibid.) along with other pioneers of human ecology, the environmental limits of growth are absolute limits due to the finite supply of stock resources on the planet. To a certain extent recycling and recovery can extend the life span of some resources. Ultimately, however when the last liter of petroleum is consumed or the last ounce of zinc removed, these resources, like thousands of species each year, will be lost forever. Remedies for some environmental problems have been found, and no doubt even more technological advances can be expected. Technical solutions cannot contend, however, with the environmental damage of increasing rates and levels of consumption. The more degraded a resource becomes, the more technologically challenging its remediation will be. Inevitably there is a time lag to recovery, and given the pressures of extravagant consumption in the North, the continuing displacement to the North of resources from the South, and increasing demographic momentum in the poor countries the limits of the planetary carrying capacity are fast approaching (Brown et al. 1989–1994). Growth economics is directly and indirectly inimical to the environment (Ophuls and Boyan 1992).

The environmental crises we are experiencing today, or are likely to experience, derive in all probability[9] from industrialization, urbanization and the denigration of rural life generally. These unanticipated crises generate an overwhelming tension in the relationship between society and nature. Such tension is unavoidable given humanity's efforts to identify Modernization with the subjugation of nature.

Growth economics, even in its most "green" garb establishes an adversarial relationship with nature. Given that place-specific conditions create economic frictions capital is compelled to either conquer, adapt to or retreat from environmental constraints, or "externalities," as they are known in the economic and regional science literatures. All social phenomena, including language, culture, economy, and infrastructure, constitute adaptations to local environmental conditions. Strip malls, dirt biking and Pepsi Cola represent the conquest of place-specific conditions. The former adapts to nature while the latter destroys nature. The great exacerbator of this state of affairs is that activities destructive to nature continue to grow in intensity, magnitude, scale and pace.

THE CONSUMPTION OF NATURE

The socio-environmental contradictions of capital expansion become even more graphic given rural flight and resulting urban sprawl encountered worldwide today. Explosive urbanization creates cities that are ferocious black holes in their consumption of resources. Their role as magnets for rural outmigrants takes a double human toll: the misery of the shantytown is added to the misery of the village whose strongest inhabitants are most likely to seek the "bright lights" of primate cities.

The problem is not just capitalism's. The other classical economic tradition, Marxism, is also vulnerable on the question of growth. The credo that "need creates right" speaks to the social limits of capitalism, the social limits of economic growth. Nonetheless, as G. Hardin writes in "What Marx Missed" (1983), Marx failed to appreciate that the "need creates right" credo is unobtainable unless the boundedness of nature is considered. Regardless of whether industrial production takes place under a capitalist or socialist mode of production, prospects for the satisfaction of human needs decrease in proportion to the size and rate of population increase—not because of an inability to produce more food[10] but due to the environmental stresses of greater energy production, resource extraction, intensified ambient pollution and changes in the climate system and biotic habitats. In other words, *future* resource needs are jeopardized because of increased consumption *now* and this will remain the case whether increased consumption derives from population growth, the uneven distribution of resources, or rising "standards of living."

The consumption of nature is common to any economic system based on growth. However, capitalism is particularly deleterious given its production of "needs" through marketing and advertising and its creation of artificial scarcity on the basis of market "logic" (Commoner 1990: 168). Simply stated, capitalism requires the consumption of nature. However, as the magnitude of Soviet "ecocide" (Feshbach 1993) has shown, centrally planned economies are no less destructive to the environment. On the other hand, even the most egalitarian society will fall victim to the environmental limits of growth unless it is informed by social ecology.[11]

RECONCEPTUALIZING DEVELOPMENT

The estrangement of society from nature and the built in necessity of market economics to subordinate the latter has led tragically to our encounter with the global problematic discussed above. Part and parcel of this estrangement is the contemporary alienation of the individual from society and of individuals from themselves.

The language of development does not adequately address today's environmental, economic and social realities. Similarly, the language of development cannot explain the failure of development since it contributes to the underlying problem—Modernization. I do not seek to refurbish development jargon. I would prefer not to use to use the terminology of development at all because, as stated by Sachs, it is necessary for us "liberate ourselves from its [development's] domination over our minds" (1992: 5). The term "development" is most correctly understood not in terms of its pretensions but in describing the historical process resulting in *global dysfunction* that has emerged from the environmental, economic and ethno-national problematic described above.

To understand the role of development in producing global dysfunction, we must acknowledge how dangerously close to the "end of nature" we have come:

> We have killed off nature—that world entirely independent of us which was here before we arrived and which encircled and supported our human society. There's still something out there, though; in the place of the old nature rears up a new 'nature' of our own devising...
>
> Simply because it bears our mark doesn't mean we can control it. This 'new' nature may not be predictably violent. It won't be predictability *anything*, and therefore it will take us a very long time to work out our relationship with it, if we ever do. The salient characteristic of this new nature is its unpredictability; just as the salient feature of the old nature was its utter dependability.... (McKibben 1989: 96)

Development is about the attempt to remake nature and its relationship to the individual and to society. We are now realizing how socially, economically, environmentally and existentially unsustainable this endeavor has proven to be. Society and its physical environment constitute an integral whole, a "single social reality" (Deutsch 1966: 41) historically defined by the evolution of the local mode of production, culture, resource use patterns and other aspects of the society-nature interface (Figure 1.2). But with global change a new reality has emerged. It is the reality of global dysfunction.

THE DEVELOPMENT OF GLOBAL DYSFUNCTION

What is referred to as development and underdevelopment are manifestations of a single historical trajectory of growth that began with the onset of the European economic expansion and the Industrial Revolution. This trajectory has produced a new global division of labor based on the economic heirarchization of societies (Dicken 1992; Knox and Agnew 1989).

While the term "development" implies an extension of available choices, lack of empowerment and social progress are the staples of life in the deprived societies of Africa, Asia and Latin America. Further, considerable segments of the populace residing in the advanced industrial countries live in similar deprivation. The argument advanced here, though, is that increasing environmental instability, social unrest and maladaptive economies are forging a set of unbearable stresses that will affect the industrialized North no less than the less developed South. The more that societies "develop," the more dysfunctional land-people relations become. This is a global phenomena affecting affluent and poor societies alike. In this context the product of the Modernization process has been the *development of global dysfunction* rather than *the advancement of societies.*

Global dysfunction can be conceived in terms of *Life-Place Displacement* and *Hypergrowth* (see Figure 1.3) involving the socioeconomic, political, cultural and ecological structures of social life. These pathologies are pervasive, structural, relational and systemic. They are chronic and persistent, cumulative and self-compounding.

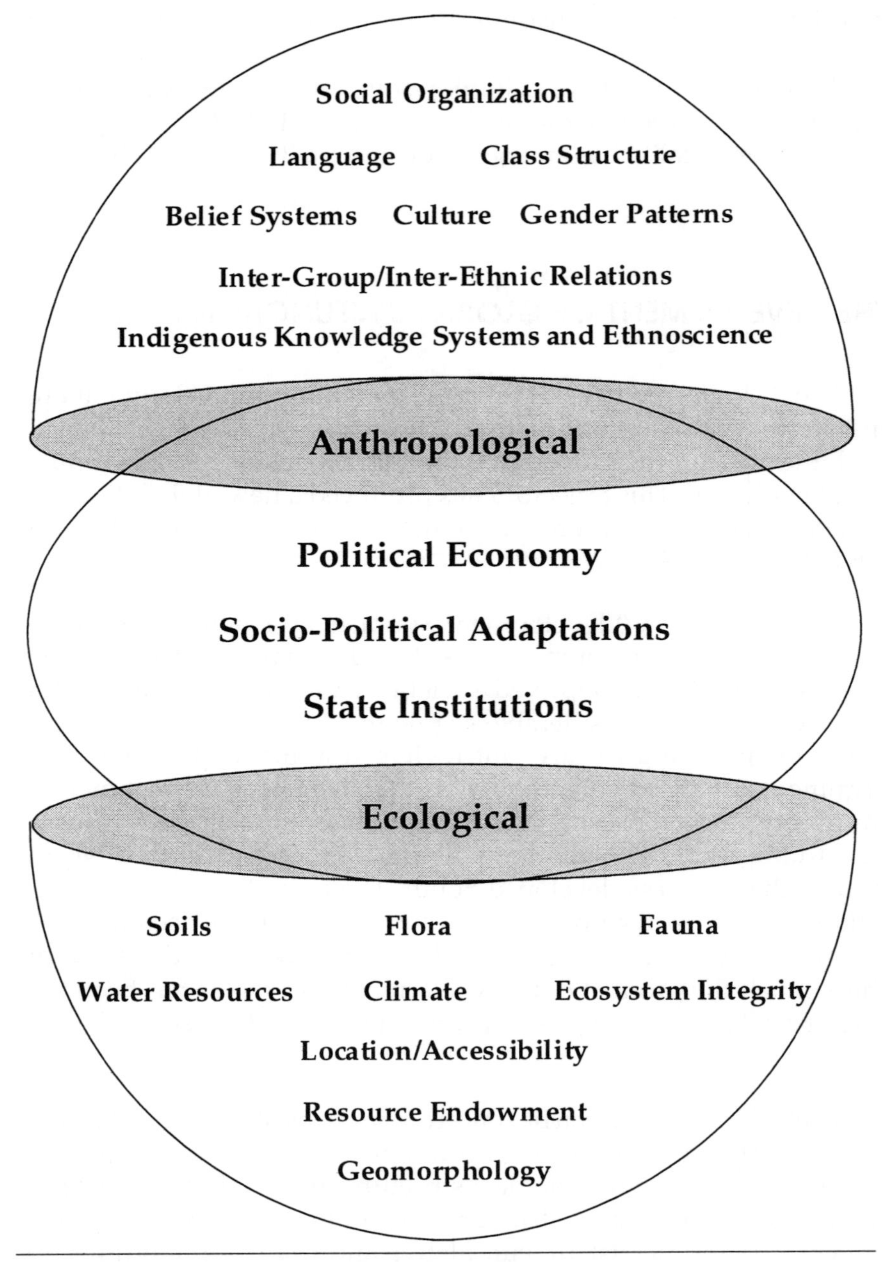

Figure 1.2 The anthropological-ecological integral.

Among the manifestations of global dysfunction are those appearing in Table 1.2. Dysfunction is defined here by a low and/or declining *potential for recovery* from natural or human-generated stress. The relative strength of land-people systems is found in their ability to resist long-lasting or irreversible damage. Irreversible effects of crises reduce the options available to society to cope with endogenous or exogenous change. Likewise, to the extent that "development" produces irreversible processes such as the death of an ecosystem (for example, Lake Baikal in Siberia) and climate change, a net loss in environmental resources available for use literally leaves the society poorer. Such loss is more immediately and dramatically felt in areas close to the source of the environmental damage, but over time it affects the ecosphere as a whole.

Many of the pathologies leading to global dysfunction originated from the ascent of industrialization and technology, particularly during the past one hundred and fifty years. These are the by-products of global economic expansion, first under imperialism, then under nationalism and corporate capitalism, of what would become a single global economy. These are global phenomena induced by growth economics and unevenness in resources, trade, technological capacity and similar forces of change. What had been life-place (Gotlieb 1992b) has been encapsulated within an expanding space economy that now comprises a unitary, although internally differentiated world-system (Dicken 1992; Johnston 1989; Johnston and Taylor 1986; Knox and Agnew 1989; Taylor 1986; Wallerstein 1982). In this system peoples and individuals are abstracted into producers and consumers, and place is replaced by economic space. Place disintegrates into an expanding space economy. Landscapes and ecosystems are literally bulldozed into shopping centers or other "developments" generated by the artificial scarcity and extravagant consumption that differentiate poor societies from wealthy ones (and rich from the poor in all societies).

An example of this process is the deforestation of tropical woodlands and their conversion into grazing land for beef cattle to be exported as meat. This transformation impoverishes the affected country by converting natural resources (forests, fauna and their

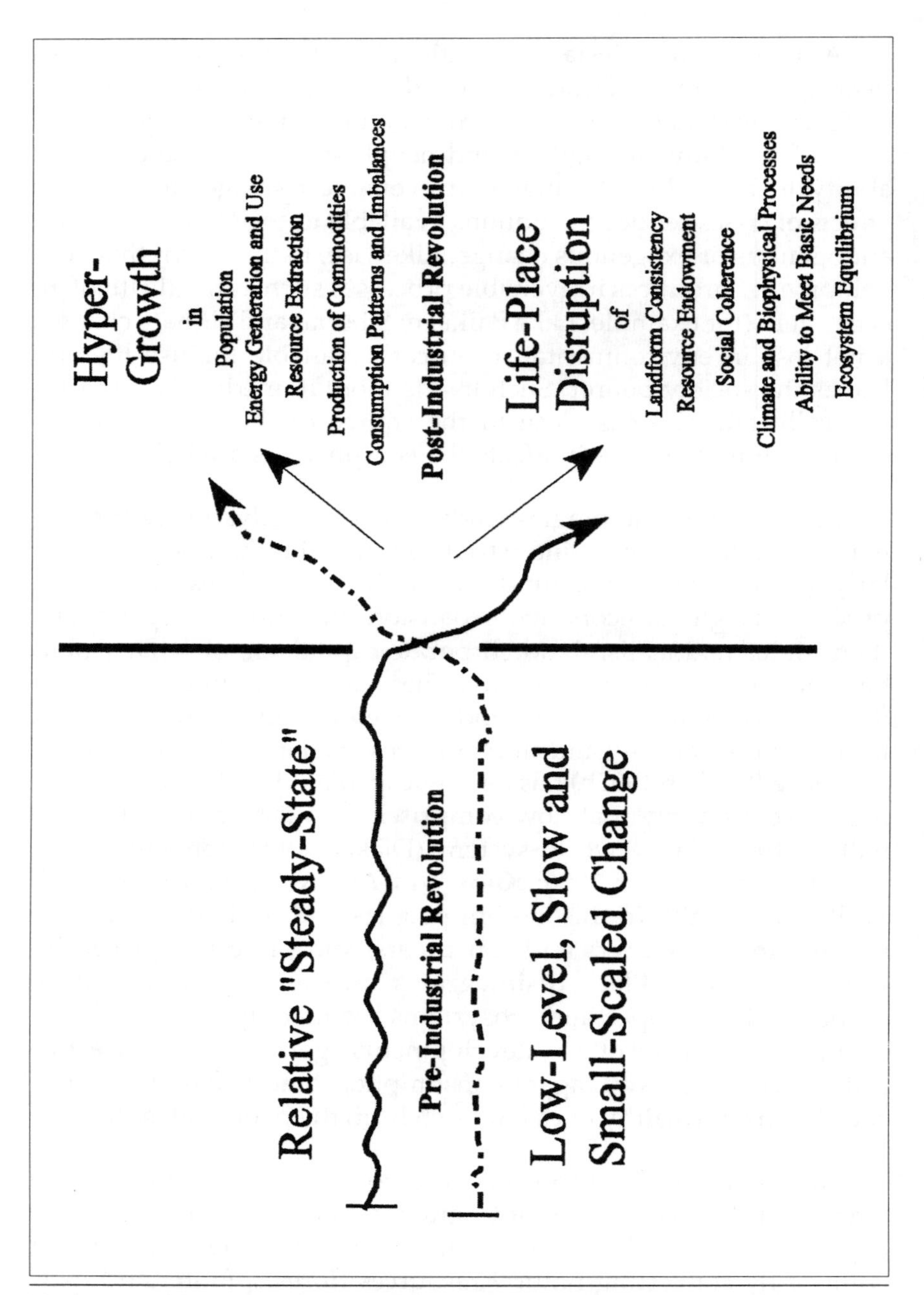

Figure 1.3 Global dysfunction: hypergrowth and life-place disruption.

Table 1.2. Manifestation of Global Dysfunction

A. Environmental Crisis/Life-Place Disruption

1. Ambient pollution
2. Contamination of water stocks
3. Accumulation of hazardous wastes
4. Fire-wood crisis
5. Land conversion and degradation
6. Ecosystem and habitat disruption
7. Reduction in biodiversity
8. Production of synthetics resistant to decomposition
9. Climate change
10. Greenhouse effect

B. Hyper-Growth in

1. Resource extraction
2. Production
3. Consumption
4. Urbanization
5. Energy generation
6. Population increase

C. Socio-Political Conflict due to

1. Mounting ethnonational unrest
2. Inequitable gender patterns
3. Religious fanaticism
4. Sense of malaise
5. The eclipse of local culture by global culture
6. Rural out-migration and urban sprawl

D. Global Economic Instability

1. Growing income disparities within and among societies
2. The foreign debt crisis
3. The inability of the "absolute poor" to meet daily subsistence needs
4. Concentration of capital and control over capital and production factors by elites and transnational corporations
5. Distortions in labor utilization (scarcity, unemployment, under-employment)
6. The failure to cost the value of environmental resources (particularly of stock resources) in production
7. Heightened speculation in capital as a commodity
8. Artificial scarcity of commodities to increase price (e.g., petroleum products)
9. Disarticulated modes of Production (e.g., formal and informal sectors)
10. Disposibility and compromised quality rather than durability in marketed goods
11. The production of "White Elephants," (large-scale, capital intensive public works that fail to meet stated objectives)

related stock resources) into artificial capital (exportable beef) that is transferred abroad and used for raw materials in the production of commodities (hamburgers); not only are these resources exported but so is their added value. Accordingly, not only has the society lost unrecoverable natural capital, but the gap between poor and wealthy states expands, resulting in both economic and ecological loss.

Similarly, converting coastal wetlands into a residential subdivision or resorts invariably leads to a reduction in biodiversity involving the temporary or permanent loss of a species that is integral to the local ecosystem. Every species occupies a niche in the system and performs an ecological function on which the ecosphere as a whole depends (Commoner 1971; Ricklefs 1993). In the absence of the species and until other species can assume its functions, the food and energy chains are destabilized. In fragile ecosystems this damage can be devastating. The costs are ecological and by extension, economic. The loss of a key plankton as a result of wetlands conversion upstream, for example, may reduce the quantity of seafood downstream. This portends dire economic ramifications for inhabitants of area villages where fishing is the dominant economic activity, as in the case of New England and Canadian fishing villages along the Atlantic coast. Further, the elimination of that one species represents the loss of part of our global genetic heritage, a heritage that has been millions of irretrievable years in the making. Such loss literally leaves humanity poorer than it had been before the application of "bulldozer technology" (Ophuls and Boyan 1992: 164).

In a state of global dysfunction, limiting the damage to the ecosphere and to group and individual identity is critical. As discussed below, the third dimension of global dysfunction, the socio-political dimension, manifests itself through social alienation, political unrest and cultural conformity rather than diversity. The loss of culture, like species loss, represents a *permanent* reduction in resources available for human progress.

If society is to succeed in surviving global dysfunction, the application of the principle of *sustainability*[12] is imperative. Concretely, the criteria of *reversibility/irreversibility*[13,14] is proposed to determine whether a social adaptation to global change is sustainable or not. Insofar as human activity produces change that is reversible (or stated more affirmatively, promotes the options available to a society for indefinite survival), it is sustainable. Adaptations that are irreversible reduce the recovery potential available to the society in proportion to the scope and gravity of the damage produced. In gauging social change, the prospects for the reversal of damage due to human agency can serve as an indicator of sustainability.

Sustainable socioeconomic adaptations are, however, a necessary but insufficient condition for arresting global dysfunction. As stated earlier, a pretension of development is its implicit claim of change originating from within the social group. The promise of development has, though, been wanting in reality. Whatever change has taken place has been exogenously imposed and most often based on the emulation of non-indigenous models of growth. Exogenous "development" is ultimately self-contradictory: there can be no surrogate in an entity's attempt to unfold from within. Hence, sustainability must be coupled with "*endogeniety*" (Kothari 1989). As used here, endogeniety refers to the achievement and/or recovery of indigenous capacity to retain control of local resources, cultural autonomy and political self-determination.

The approach employed in this study synthesizes various perspectives on development and the environment. This synthesis provides tools for understanding the economic connections that link life-place to higher scales in the global space economy. The resulting social ecology eschews any idyllic or utopian visions of what the world *should* be like, nor does it entertain a romantic view of the pre-industrial past which is, in any case, unrecoverable (Weitz 1986). The recovery of whatever endogenous capacities, human and non-human that are still viable to a particular place is the goal of social ecology. This entails the restoration and autonomy of local cultures that Verhelst (1990) and others have shown to be vital to progressive social change.

SOCIAL ECOLOGY AGAINST DEVELOPMENT

The social ecology approach advocated here identifies with the political ecology school as defined by Blaikie and Brookfield: "The phrase 'political ecology' combines the concerns of ecology and broadly defined political economy. Together this encompasses the constantly shifting dialectic between society and land-based resources, and within classes and groups with society itself" (Blaikie and Brookfield 1987: 17). Other important works in the political ecology tradition include those of M. Watts (1983a, b) and T. Bassett (1988). Social ecology employs analytic tools used in political ecology and other approaches such as eco-development (Bartelmus 1986), ecosystem (Ellen 1982), territorial development (Weaver 1981), basic needs (Cerena 1985; Stohr 1981; Streeten, et al. 1981), human-scale development (Max-Neef 1991), and ecological economics (Costanza 1991), among others. It is not only a framework for evaluation and critique but also for determining sustainable endogenous responses to global dysfunction.

Natural conditions offer a multiplicity of developmental possibilities for—and constraints on—human agency as the primary factor shaping society (Geertz 1963: 1–2). The growing recognition that humanity has become estranged from its environment and that the post-colonial state contributes to this estrangement has caused the reexamination of the modern state system. As elaborated below, the modern state negates nature. Underlying the butchery of inter-state conflicts is the disintegration of the society-nature integral. The latter is dynamic and evolves over time. Given its Modernization mission, the post-colonial state relies on the elimination of nature and of ethnic identity. An alternative to the post-colonial state framework is badly needed. Therefore, regionally-based endogenous recovery rather than state-centered development and aggregate economic growth is proposed here.

The social ecology approach posits a territorial concept of the region based on current anthropological, ecological and economic continuities. The recovery of areas where anthropological and ecological continuities and coherence still exist requires socioeconomic change that is endogenously achieved. Although the life-place of traditional peoples have been irreparably changed and a new set of adaptations is required to contend with the altered landscape that has replaced it, the possibilities for endogenous recovery of our collective and individual selves is much more feasible than the artifice of development.

The vessel for endogenous recovery is termed the *Endogenous Recovery Region,* a territorial region where traditional anthropological adaptations to the environment, most notably the mode of production remain relevant even if distorted. This is in contradistinction to the functional region of space economies as interpreted by positivist regional science (Max-Neef 1991) where place-specific conditions are modified to reduce impediments to capital expansion.

Endogenous Recovery Regions may be conterminous with, extend beyond, or partially comprise the borders of existing states. Where societies and ecosystems are fragmented, post-colonial states may have to be territorially reconfigured—for the sake of endogenous recovery and self-determination. Where such fragmentation is less profound and consensus can be reached between ethnic groups inhabiting different niches of the same social environments, existing state borders may remain intact.

The approach articulated below does not advocate closed and uniform societies devoid of internal ethnic diversity, nor does it envision societies where the majority population is politically or economically privileged and dominant over minorities. Democracy and popular participation are viewed as inextricable from sustainable endogeniety and socio-political self-determination. Only on this basis can the repair of the broken peoples and broken lands of the post-colonial world take place.

THE ORGANIZATION OF THIS VOLUME

The first part of this book explores the dimensions of *global dysfunction* (i.e., the multiple economic, social, environmental and political maladies that threaten the survival of the ecosphere (Chapter 2). Chapter 3 interprets the lessons to be learned from global dysfunction in terms of the relationship between society and nature. In Chapter 4 guidelines are proposed for identifying the appropriate socio-spatial framework in which sustainable endogeniety may be applied (*the Endogenous Recovery Region*). A summary statement of the social ecology perspective is presented in Chapter 5. Chapter 6 applies the Endogenous Recovery Region to a concrete case, the Kurds and Kurdistan.

The final chapter constitutes a proposal for further research and policy changes concerning endogenous recovery in a world where global economic and environmental processes increasingly denigrate the importance of place.

NOTES

1. The start of contemporary "development" is generally associated with the Bretton Woods Conference held in New Hampshire in 1944. The International Bank for Reconstruction and Development (the World Bank) and the International Monetary Fund are products of the Conference. Conceived as instruments for the reconstruction of Europe's war-torn economies, these agencies were to assist post-colonial states under the assumption that while these states lacked the capital and infrastructure that the Europeans had, pushing them along the road to development was, at least structurally, no different than the road to recovery Europe followed.

2. See Wisner (1988: 51–52) for a similar articulation of the Modernization paradigm and Barnett's (1989) excellent discussion of Modernization and its detractors.

3. The need to recraft our view of social change is addressed by D.W. Orr's concept of Ecological Literacy (Orr 1992).

4. The manner in which post-colonial states were created and their socio-spatial composition is highly representative of Modernization's assault on traditional social formations and their relationship to the physical environment around them. This is discussed more fully in succeeding chapters.

5. W.W. Rostow's *The Stages of Economic Growth* is perhaps the best known work on stage theories of growth. It is almost universally accepted that stage-appropriate strategies for development is a formula for the emulation of the advanced industrial economies by the low-income societies in the South.

6. For a fuller exposition of the life-place concept see Gotlieb (1992b).

7. See José Maria Sbert's (1993) essay on the subject.

8. See Rees (1990: 51) on this point.

9. Without entering into the debate here, it is important to note that a plausible case can be made concerning non-human origins of recent environmental instability. Formal epistemological concerns notwithstanding, whether one believes that growth activities might, probably do or definitely do have a negative impact on the environment, the implications remain the same. We are treading on risky ground.

10. Our technical ability to produce prodigious quantities of crops was demonstrated by the successes of the "Green Revolution." However, no less important are the inherent drawbacks of High Yielding Varieties in terms of water consumption, fertilizer intensity and toxicity, energy requirements and other inputs. The social consequences of the Green Revolution have actually exacerbated the socioeconomic gap between rich and poor since HYVs require inputs that are far beyond the means of smallholders. While there have been improvements in the use of HYVs and the infrastructure necessary to support them, the technology of the Green Revolution must be applied with great care. See E.C. Wolf (1986) for further background on the subject.

11. The term "social ecology" as used in this essay is informed by, but distinct from, the meaning assumed by Murray Bookchin's prolific and pathbreaking contributions to the human ecology literature.

12. For discussions of the evolution of sustainability as a concept see Costanza (1991), Orr (1992), MacNeill (1990) and Ruckelshaus (1990), or Friedmann (1992).

13. See Barbara Ward's (1973) discussion of irreversible environmental change as the primary threat posed by ambient pollution.

14. Irreversibility is not an absolute attribute in the sense that it has different meanings over time. Geological time is the temporal scale in which geological changes takes place. It is a very long period of time ranging from thousands to millions of years. Resources such as fossil fuels that could theoretically be renewed given the right combination of conditions over geological time are effectively unrecoverable over human time (the period spanning three or four generations).

CHAPTER 2

GLOBAL DYSFUNCTION: MULTIPLE PROBLEMS, ONE PROBLEMATIC

Recent studies (World Bank 1988, 1989, 1990; WRI 1992: 15; WRI, UNDP and UNEP 1992) indicate that socioeconomic conditions worsened in the low-income countries throughout the 1980s. Little improvement had been recorded despite the broad array of development policies applied:

> Poverty in the developing countries is on the rise....Since 1980 matters have turned from bad to worse: economic growth rates have slowed, real wages have dropped, and growth in employment has faltered in most developing countries. Precipitous declines in commodity prices have cut rural incomes, and governments have reduced their spending on social services (World Bank 1988: 4).

No less grave than retrograde development are its correlates in the form of global and local environmental degradation and social, particularly ethnic, conflicts. This study suggests that these sets of phenomena are inextricable and are symptomatic of global dysfunction. To assess their synergistic effect, it is useful to survey each set of maladies.

DECLINE IN THIRD WORLD DEVELOPMENT

The deterioration in the status of the low-income countries over the last fifteen years is substantiated by numerous indicators:

> Rising poverty rates in Africa, Latin America and parts of Asia appear to have swamped poverty reductions in India and China, with the result that in 1989 approximately 1.2 billion people lived in absolute poverty. In part because of population growth, that total is much larger than ever before. Perhaps more significant is what this means overall. World Bank figures suggest that the global poverty rate stood at 22.3 percent in 1980, after declining steadily but gradually since mid-century. Our own estimate of 1.2 billion people translates to a rate of 23.4 percent. During the eighties, in other words, the global poverty rate not only stopped falling, it rose—despite substantial reductions in the number of impoverished in the two most populous countries on earth [China and India] (Durning 1989).

These figures are particularly troubling given that poverty began inauspiciously during the 1980s, only to worsen. The Report of the Independent Commission on International Development (the Brandt Commission) published in 1980 describes then current conditions in depth. Many of these conditions have since deteriorated.

Given the dismal conditions existing at the beginning of the 1980s, the decline over the course of the decade as described above is especially frightful. Although real incomes rose from 22 to 37 percent in the East Asian countries during the period 1960–1989, the

Latin American and Sub-Saharan African regions actually saw a reduction in their real incomes by six and five percent, respectively. In the 1980s the fall in GDP in Latin America and Sub-Saharan Africa was aggravated by reductions in investments. For the developing countries as a whole, GDP dropped in 1980–89 from 5.9 to 4.3 percent per annum relative to the 1965-80 period. These same regions saw a decline in the terms of trade during the eighties by as much as 13 and 15 percent (World Bank 1990: 7–15).

Aggregate economic data are too removed from field realities to convey the full depth of rural poverty. Often data is only remotely representative of the situation in rural areas, although they are presented as characteristic of the society as a whole. Aggregate data is highly problematic and limited by its design, means of collection and the accessibility of the population being characterized. Analysis of aggregate trends is often little else than the manipulation of indicators of central tendency. Accordingly, while the profile presented may approximate that of the major urban areas, it may seriously misrepresent the situation elsewhere in the country.

However, the problems with development data extend far deeper than the distorted profile projected by aggregate analysis. The difficulty with using such information is a fundamental one involving the concept and mission of development, specifically of economic growth as the standard used to measure such change. The fixation on econometric indicators has "impoverished" development (Esteva 1992) by diverting attention from structural problems. Among such problems are (a) the failure of positivist economics to insure that the spread of benefits reaches those most in need of them, (b) its failure to factor ecological costs (e.g., the costs of depleting stock resources) into social debts, (c) the view that social welfare embodies products rather than rights, and (d) conventional economics' inability to assess endogenous capacity for cultural, institutional, political and ecological recovery.

Accordingly, the mixed picture of economic growth during the last decade as described by the World Bank and other sources does not fully reflect socioeconomic realities. Social welfare indicators suggests that development was much more problematic than would appear to be the case in reviewing economic data alone. For example, food production per capita, particularly of grains, and soybean and animal proteins, have either remained static or declined (Brown, Flavin and Kane 1992: 17–18). While infant mortality rates have dropped, maternal mortality rates in Third World countries remain higher by several orders of magnitude than those in the societies of advanced market economies (ibid.: 19, 112–113). Additionally, the income gap between classes in numerous poor countries actually rose during the eighties (ibid.: 19, 110–111). Moreover, the gender patterns of poverty are conspicuous and show women as being particularly hard-hit by poverty and environmental stress (Kelkar and Nathan 1991; Rocheleau 1991; Shiva 1991, 1989; WRI 1992: 31).

ENVIRONMENTALLY UNSUSTAINABLE DEVELOPMENT

V. Shiva's elegiac *Staying Alive: Women, Ecology and Development* speaks not only to the double burden of women in underdeveloped countries but to the state of the environment as well: "...The earth is rapidly dying: her forests are dying, her soils are dying, her waters are dying, her air is dying..." (1989: XV).

With the issuance in 1992 of the *World Development Report,* the World Bank acknowledged that underdevelopment and environmental degradation are inseparable. For example, according to the *Report* (1992: 1–93) the contamination of water sources has left one billion people without access to clean potable water, a situation resulting in death, disease, and at least temporary incapacitation for hundreds of millions of people each year. While urban populations (and rural areas in the pollution shadow of cities) suffer from the effects of industrial air pollution, 400 to 700 million people—mainly women and children—in poor rural areas have their health compromised by the inhalation of smoky, indoor air deriving from the burning of organic fuels (wood, charcoal, dung) for cooking and heating. The multiple deleterious effects of the firewood crisis on health and the environment has become a major concern of development and environmental specialists (Figures 2.1 and 2.2) (WCED 1987: 189–192).

At a time when massive population growth and concomitant food demands are on the rise, soil productivity has dropped-off markedly as a result of soil erosion, salinization, waterlogging and especially deforestation (Brown 1990; Brown and Wolf 1985; Little, Horowitz and Nyerges 1987). Consequently, the amount of land that can be farmed or used for grazing has steadily dropped. Deterioration in soil quality affects the poorest of the poor since the latter are found predominantly in rural areas (Blaike and Brookfield 1987; Ophuls and Boyan 1992: 49–51). With the loss of tropical forests occurring at a rate of at least 0.9 percent annually the health and livelihood of 140 million people are directly affected. The indirect results extend even further. Consequently, the causes of the problem and the numbers of people it affects are continuously compounded.

Deforestation is not only a major catalyst of soil erosion, but it contributes to climatic disorder such as irregular precipitation cycles (Iltis 1985). Further, it results in loss of biodiversity (including species that will never have been documented (Shiva et al. 1991; Wilson 1990), in landslides, and in atmospheric damage and numerous other interrelated problems on an expanding scale.

Since poor societies are overwhelmingly rural societies (World Bank 1992), they depend directly on what they are able to draw from the land. Their poverty, however, aggravates and is aggravated by environmental degradation (WRI 1992–1993: 29–40). The industrialization of countries with advanced market economies has directly or indirectly spoiled and depleted air, water and land resources on a global scale, yet it is the peoples of the South rather than those of the developed countries who suffer the most from compromised natural resources (ibid.: 4). Since these peoples directly depend on place-specific combinations of soil, fauna and flora, water, and climate, environmental degradation translates into scarcity of those basic goods and tools that have been used by indigenous peoples to sustain their societies for centuries (Redclift 1987: 150–9).

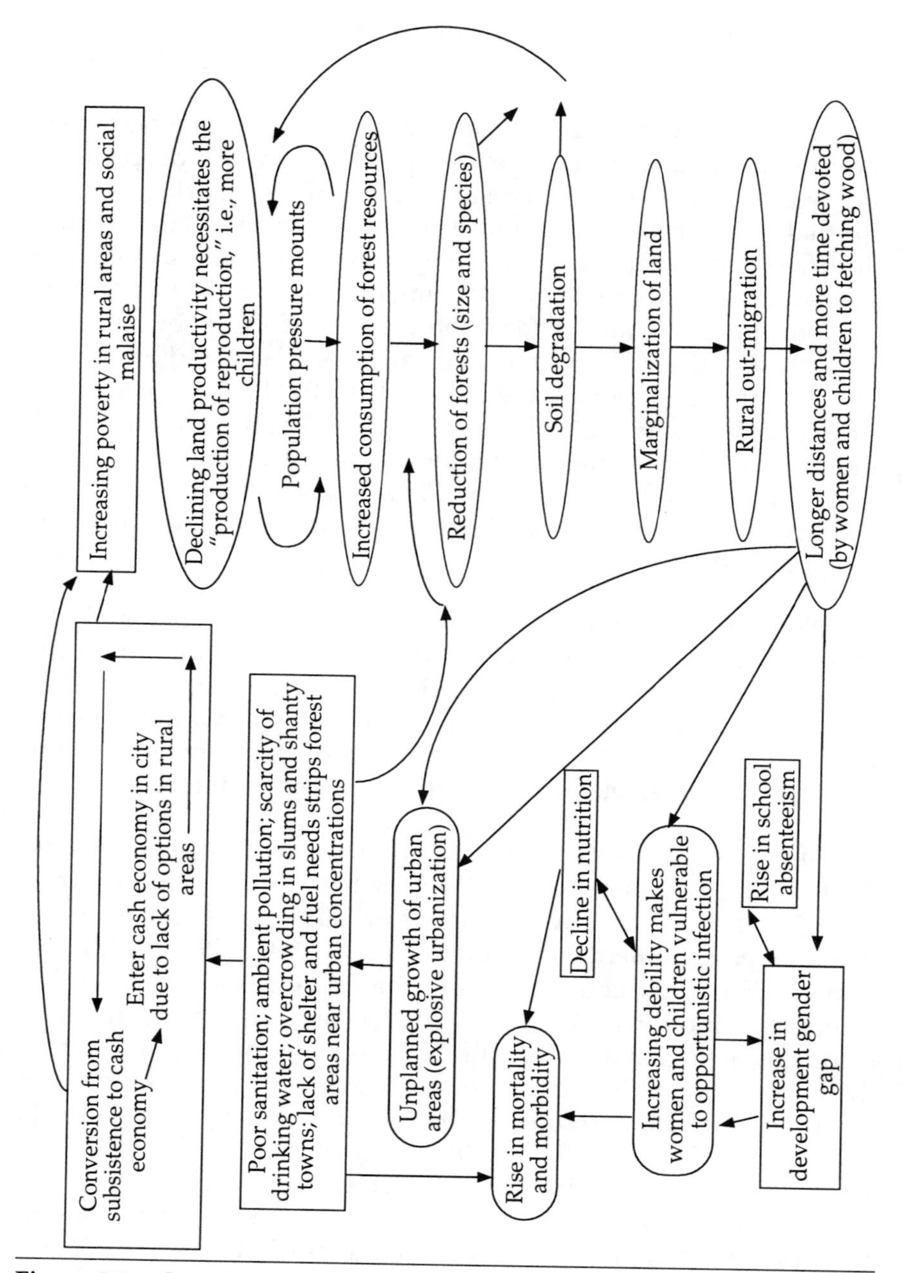

Figure 2.1. Cumulative circular cause/effect of the fuelwood crisis.

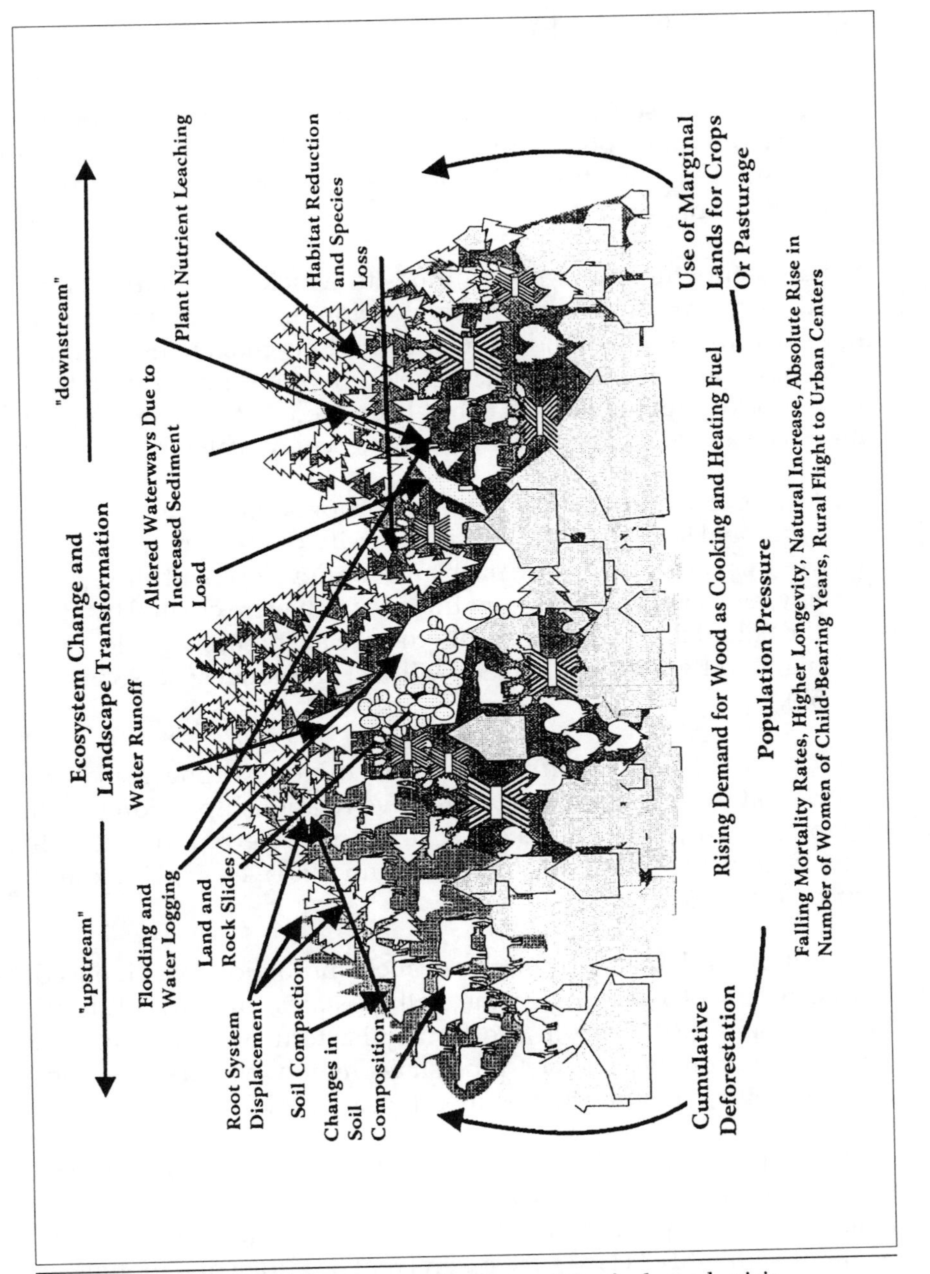

Figure 2.2. Multiple environmental effects of the fuelwood crisis.

THE ENDURANCE OF ETHNIC CONFLICT

Unremitting social unrest and ethnic discord have become a mainstay of contemporary life (Clay 1992). These conflicts are as sure a sign of dysfunctional development as poverty and environmental degradation.

Persistent ethnic[1] unrest continues to wrack Second and Third World states. At a time when many developing countries are entering their fifth decade of independence, serious challenges to political stability and socioeconomic development are posed by ethnic dissidence or regionalism. This has been a curse that has plagued Third World peoples since their independence. As stated by D. Seers:

> [A] topic widely ignored, indeed suppressed, by all conventional schools is the ethnic diversity of many nations; yet this raises very basic problems for development strategy, especially in the non-nations...which were created by the colonial powers (probably the greatest crime of colonialism, because the effects will last the longest) (1983: 57).

Ethnic unrest has been on the rise, and there is strong reason to believe that there is a mutually causative relationship linking such phenomena with the process of social change referred to as development/underdevelopment (Hirschmann 1981: 49–58; Knox and Agnew 1989: 366).

Multiple and diverse examples of such imbroglios are evidenced by the recent bloodletting in the Balkan states, Bulgaria, Russia, Romania and elsewhere in the former Eastern Bloc. The strong, centralized tyranny of Soviet-type states merely obscured the subterranean ethnic fissures that defined those societies.

Concerning Asia, H. Tinker relates that with the possible exception of Japan, China and Korea, Asian countries have "massive majority/minority difficulties that inhibit the formation of a sense of national identity" (1981). He asserts that the "idea of nationalism remains...the cement that holds together the state...though it is increasingly clear that the cement is mainly sand." He concludes that the Third World nation-state today "does little or nothing to help the Asians solve their most immediate problem: that of institutionalizing the relations of communities and groups, whom geography has made neighbors but who possess no real feeling of a common identity" (ibid.).

In response to these and other problems, A. Hughes declares that what are called nation-states in the Third World are nothing more than "colonial states" (1981). The system found in the South today is one wherein formerly free and cogent social formations are divided by the borders of post-colonial states which are, in effect, states without nations or states with multinational societies so heterogeneous as to render them ungovernable. This dismemberment of Third World peoples and their life-places has effectively prevented the achievement of comprehensive self-determination (Cobbah 1988) and has impeded development (Seers 1983). Perhaps formulas could be found to render inter-group relations more harmonious but the lack of congruence between nation and state is not the sole problem. The states themselves lack historical continuity and range from gargantuan to microscopic in scale. They must contend with heterogeneous populations varying in size from the low hundreds of thousands to the high hundreds of millions. The state boundaries encasing these populations were drawn without consideration of carrying capacity, resource endowment or the integrity of the ecosystems on which these societies depend and of which they are a part.

Ethnic unrest in Latin America is less central a question than in Africa and Asia owing to the relative homogeneity of populations—due to the success of European conquistadors in exterminating the native populations of the Americas. Still, in the Andean countries, Amazonia, Guyana, Nicaragua, Guatemala, and Mexico, the aboriginal Amerindian populations struggle to preserve their societies and historical habitats. They, along with Kurds, Tibetans, Karens, Sumatrans and scores of other ethnonational groups, persist in demanding the right to control their own collective lives.

SOCIO-SPATIAL REALITIES OF TERRITORY

Why is the ethnic dimension so significant? Because in rural societies, it powerfully represents the lines of conflict over increasingly scarce resources. Attempts to "bulldoze" and submerge ethnic identification as part of the Modernization enterprise consistently encounters resistance. An analogy is found in modern technology applied to maximize crop yields. Short-term yields may rise, but the damage done to soil quality and fertility is incapable of providing food to an increasing population base.

Just as the maximization of yields is unsustainable in the long run, so is the "cultural engineering" engendered by Modernization. Technological systems to industrialize agriculture, like social mechanisms aimed at creating modern citizens, result from the objectification and manipulation of variables. Objectifying society and nature fail to deal with deeper structures and processes such as the integrity of society-nature relations. Indifference to the socio-spatial characteristics of place and their importance to endogenous social change is a major determinant and symbol of global dysfunction. The denigration of life-place through Modernization impacts on the social identity and existential support systems of those living there. Moreover, traditional land use patterns and indigenous knowledge systems often reflect time-tested, sustainable agricultural practices that are increasingly shown to be highly adaptive to local conditions.

Place-specificity is impressed both on the anthropological and environmental aspects of the society-nature integral. Society represents human adaptation to the environmental "given" of the integral. When social relations or ecological conditions are altered, the effects are circular and compounding across the society-nature interface. In many, if not most cases, ethnic unrest is shown to represent a profound disharmony between neighboring groups responding to the stresses of an altered and damaged environment.

Problems of State

A considerable body of research exists with respect to government policies and state regimes; however, a review of the literature indicates that critiques of the post-colonial state in terms of social coherence, geographic and ecological integrity and historical continuity are few in number. The Third World nation-state is generally taken as a given, and emphasis is placed much more on the policies adopted by a particular state regime than on the socio-spatial configuration of the polity it governs (Hellen 1968). As described by D. Ronen:

> ...if there is anything common to most (if not all) of the social science literature on development, Modernization, self-determination, and ethnic conflict, it is the acceptance of the existing framework of states and the state system (1979: 18).

Third World liberation movements have emphasized the struggle for state power as the sine qua non of decolonization, and decolonization is interpreted as entailing the transfer of power from imperial to indigenous regimes. Ironically, the territory of the previous colonial entity in the form of the post-colonial state has been regarded as the appropriate vehicle for the development of Third World societies—notwithstanding the social-territorial malformation inherent to these entities, which were created to foster the transfer of resources from South to North for the benefit of the imperial powers. Decolonization confers legitimacy on the regimes governing post-colonial states. These regimes are almost universally judged on the basis of their success in "Modernizing" the people residing in formerly colonized space and over which they rule (Mazrui 1972; Myrdal 1968). Yet, the underlying realities are such that the "development" of such states is akin to trying to force a round peg into a square hole.

Modernization

The Modernizationist creed is described by Ben Wisner in the following terms:

> Modernization theory defines development as the erection of a modern state and market structure. (The state socialist variant modifies the form of the state and market, but is identical in spirit.) In the view of the Modernizers, proper states and markets should not (and normally do not) allow hunger, disease, illiteracy or hopelessness. Modern social life is supposed to be about the sublimation of regional, cultural, gender and class differences to a greater, harmonizing whole, not their free expression. The purpose of the state is to unify disparate interests, while the market impersonally levels out differences and democratic institutions channel and institutionalize the atavistic energy inherent in regionalism, tribalism, class struggle and sexual politics (1988: 51–52).

In the post-World War II period development economics has been dominated by the Modernization approach, particularly market-oriented ones (Rostow 1960; Kuznets 1965; Hirschmann 1981, 1958; Little 1982). These approaches factor out place and environment from the development equation. Modernization models place emphasis on aggregate, i.e., state-level economic growth. Advocates of this approach presume that those modalities which led to the development of the market economies of the North could and should be adopted for use in the South; this has been as true of most Marxist traditions as capitalist ones (Redclift 1987: 45–51).

Market-based Modernization models of development are often premised on the notion that economic growth unfolds everywhere through uniform processes. This thesis found its best-known formulation in W.W. Rostow (1960) *The Stages of Economic Growth.* Rostow's and similar work by others have been indicted, however, as being intrinsically foreign and inimical to Third World contexts

(Amin 1990; Cobbah 1988; Friedmann and Douglass 1981; Kedourie 1970; Sachs 1991; Verhelst 1989). These perspectives aver that there is one universal path to development—that taken by the advanced market economies of the North.

The dominance of Modernization in social theory has been challenged by new treatments of social theory. For example, J. Agnew questions the abstractionism inherent to Modernization theories by stressing the significance of place-specificity on social life and as the building block of global processes:

> Political behavior is formed in a nested hierarchy of socio-territorial contexts. Places or locales are the "lowest order" of these. In Modernization theories this level is viewed as significant only in traditional or parochial societies and not in "modern" ones.... "Deregionalization" and the declining significance of place are likewise major themes in the political geography and sociological literatures on "nation building" and the effects of the mass media on political behavior...To insist on the continued importance of place, however, is not to deny that processes beyond the locality have become important determinants of what happens in places. But it is still in places that people's lives are lived, economic interests are defined, information from local and extra-local sources is interpreted and takes on meaning, and political discussions are carried on (1982).

Increasing numbers of theorists now view the place-specific conditions of societies as being of fundamental importance to the context of development (Agnew 1984; Amin 1990; Friedmann and Douglass 1981; Friedmann and Forest 1988; Knight 1982; Peet 1986; Seers 1983). In describing the depth of ethnic attachment, C. Enloe argues that if Modernization conflicts with tradition and social autonomy, then perhaps it is Modernization rather than group solidarity that should be sacrificed (1973: 274).

The Limits to State-Centered Modernization

T.G. Verhelst in *No Life Without Roots: Culture and Development* questions the very legitimacy of the post-colonial state based on the violence it inflicts on Third World societies:

> ...Nowadays, it has become obvious that most Africans hardly recognize themselves in the states their colonizers have bequeathed to them. Since the state sees itself as the driving force of development, the latter consequently finds itself profoundly handicapped. The frequency of coups d'etat reveals not only the behind-the-scenes intrigues of neo-colonialism, but also the shallowness of the regimes' roots in society, the unsuitable nature of their methods of government and the very nature of their power.
>
> [T]he post-colonial "state-idolatry" is equalled only by the profound absence of legitimacy of the authorities. An artificial entity, from the points of view of both is frontiers and its history, the African state, far from being the product of a long and spontaneous process of nation-building, exists in itself, and very often for itself and for the bourgeoisie which has taken control of it. The people are elsewhere and define themselves by a sub- or trans-state identity....The state imposes centralization and homogenization, ethnic groups demand the rights to their differences and autonomy (1989: 38).

An integral part of the social "glue" that kept traditional societies whole is ethnicity. Ethnic identity appears to be a central part of all social formations: "[E]thnicity, as one type of primordial assumption about the nature of human identity, can be found in all types of societies, industrial as well as nonindustrial" (Keyes 1981: 27). By primordial, Keyes refers to descent, that is, social rather than genetic descent; it is a learned identity and is culturally rather than biologically transmitted (ibid.: 5–28).

Efforts to treat social formations as content-less variables in an abstract calculus of development are difficult, if not impossible, to successfully implement. They generate global dysfunction, not constructive social change.

"DEVELOPMENT" AND GLOBAL DYSFUNCTIONAL

The evolution of dysfunctional development originated in the North where it profoundly transformed European societies, particularly through urbanization and the myriad economic and ecological changes brought by industrialization. Concurrently, European imperialism subverted indigenous ways of life and modes of production in Africa, Asia, Latin America and the Caribbean. Imperialist penetration and subterfuge profoundly altered indigenous societies and led to the gerrymandering of life-places. Hyper-growth in population and resource extraction occurred. Throughout this century, especially in the period after the World War II, decolonization engendered the collapse of formal control over colonized territories. However, while scores of new states were created, most of them were congruent with previous colonial borders. The transfer of power from colonial to indigenous regimes—however necessary this may be to self-determination—has not led to structural decolonization. Entrenched poverty, ecological destruction, social unrest, unevenly distributed development burdens and benefits, irregular changes in government, and authoritarianism practiced by extroverted elites continue to plague these countries.

The response to underdevelopment in the South began in earnest at the end of World War II when the former colonies became independent. At that time "experts" from the advanced capitalist countries as well as the Eastern Bloc prescribed growth-oriented solutions aimed at emulating the "successful" Modernization of the advanced market economies (Barnett 1989). Such emulation has proven to be unworkable in most Third World contexts. In searching for an alternative to development economic, theorists focused their critiques on the programmatic content informing most development policy, i.e., Modernization. Such theorists have noted that growth economics, Modernization theory and prescriptive planning generally have had deleterious rather than constructive effects (Becker and Egler 1992; Escobar 1992; Independent Commission on International Development Issues 1986(80)) and that alternative development programs were needed (Friedmann 1992).

However, critiques of development programs do not entirely explain why development activities have proven to be so universally dysfunctional. A factor appears to be missing, which has to do with accounting for the socio-spatial *framework* in which development takes place, *not just its program*. In particular, I believe that the life-place disruption created under colonialism and perpetuated in the post-colonial state is overwhelming and will defy any development regime and technology applied to it. Without a critique of the socio-spatial focus of development, i.e., the post-colonial state, the Third World problematic cannot be fully understood. Even if appropriate programmatic solutions for constructive change could be achieved, any enterprise aimed at developing the societies assembled by colonialism and trapped by the post-colonial state structures will remain dysfunctional.

The structures and institutions of such states remain oriented to an evolving world-system of capital expansion that is as debilitating to Third World peoples as direct colonialism was. Development activity must be spatially and socially refocused to conform with socioeconomic, political and territorial realities distorted yet still vibrant under colonialism and in the post-colonial state. Whether referred to as an ecoregion (Bartelmus 1986: 45), the spatial focus of regional political ecology (Blaikie and Brookfield 1987: 17), or as the Endogenous Recovery Region as proposed here, an entity based on the ecological and anthropological continuities obscured by colonialism and the post-colonial state is imperative. In this light, development activity based on state-centered policy is crucial.

To summarize, global dysfunction as described above derives from Modernizationist development. Social ecology must overcome two basic deficiencies in our understanding of why development has failed. The first of these is the wholesale disruption of life-places entailed by development. The second deficiency is the failure to realize that hypergrowth is a global malady that threatens the entire ecosphere. The trajectory of Modernization has led to environmental, economic and political dysfunction. Discarding the development paradigm is indispensable to global recovery. The program of social ecology is offered as an alternative, one geared toward sustainable, endogenous recovery rather than the hypergrowth of global capital implicit to Modernization.

NOTES

1. Concerning the sociological definition of ethnicity or ethnic group, Shibutani and Kwan's description is adopted:

   > An ethnic group consists of people who conceive of themselves as being of a kind. They are united by emotional bonds and concerned with the preservation of their type. With very few exceptions they speak the same language, or their speech is at least intelligible to each other, and they share a common cultural heritage. Since those who form such units are usually endogamous, they tend to look alike. Far more important, however, is their *belief* that they are of common descent, a belief usually supported by myths or a partly fictitious history. A people do not necessarily constitute a nation; although men who regard themselves as being of a kind tend to move in that direction, they are not necessarily united under a single government (1966: 40–41).

   While the above definition speaks to the social component of ethnicity, additional dimensions—territorial and economic, among others—are viewed as inextricable, as will be elaborated later.

CHAPTER 3

THE SOCIETY-NATURE RELATIONSHIP

This chapter explores a broad body of literature—human geography, human ecology, ecological anthropology, cultural ecology, political ecology and ecodevelopment, among others—in an effort to gain insight into the relationship between societies and the lands they inhabit. An understanding of this relationship and the disciplines dealing with it are critical to an explanation of the development of global dysfunction.

At the heart of global dysfunction is the shearing of the anthropological from the ecological. My contention is that the society-nature relationship has been disrupted by a "technosphere" (Toffler 1989) that transforms both.[1] Explanations for this disruption are varied since diverse approaches offer different conceptualizations of land-people relations. Despite this diversity they share the view that the relationship between societies and their environments is an integral and symbiotic one.

EXTREMES DEFINING THE SPECTRUM OF SYMBIOTIC HUMAN-LAND RELATIONS

Two extreme and opposite positions define the continuum of paradigms premised on the principle of society-environment symbiosis (see Figure 3.1). Lying outside this continuum are approaches that view either society or environment as predominant over the other. The commonalities of the broad array of perspectives described below is the belief that the society-nature relations are dynamic and mutually formative.

The two extremes falling outside the continuum of land-people symbiosis assert that the relationship between nature and society is a zero-sum game: one dominates the other. At one extreme is environmental determinism of Montesquieu, Ritter, Ratzel and Kjellen, and Haushofer (Kasperson and Minghi 1969). These theorists postulate an organismic model of the land-people relationship whereby the relative advantages and disadvantages conferred by nature on different territories render societies inhabiting these lands relatively superior or inferior to other peoples. The opposite extreme of the spectrum is defined by perspectives that measure progress in terms of extending human control over nature: changes in landscape.

Figure 3.1 The polarity of approaches to society-nature relations.

Both the environmental determinism and the extreme human agency perspectives are flawed since they view society and environment as separate spheres. Both extremes violate a basic principle that is widely espoused among the approaches presented below: The society-environment relationship is a unitary one; people and land are an integral entity. For analytical purposes they can be studied separately. However, while the dynamic between society and environment might be variously interpreted, meaningful knowledge of the human-land relationship is based on a recognition of their mutuality and synergy.

THE SPECTRUM OF SYMBIOTIC APPROACHES TO SOCIETY-ENVIRONMENT RELATIONS

The diverse approaches to society-environment relations share the perspective that society is an essential part of (or component in) nature.

Human Ecology

Human ecology constitutes that pole of the symbiotic spectrum that is adjacent to—but decidedly distinct from—the environmental determinism extreme. As D. Hardesty writes: "Today the theme of environmental determinism has been largely replaced by the emergence of man-environmental models that assign environment a 'limiting' but uncreative role or that recognize complex mutual interaction" (1977: 3). Concepts such as population, habitat, ecosystem, energetics and system equilibrium are frequently considered. The notion of human adaptation to an ecosystem is emphasized.

Culture and Nature

A. Hawley's views are representative of the human ecology school:

> The meaning of human ecology may...be stated as a paradigm composed of three propositions. They are: (1) adaptation proceeds through the formation of interdependencies among the members of a population; (2) system development continues, ceteris paribus, to the

> maximum size and complexity afforded by the technology for transportation and communication possessed by the population; and (3) system development is resumed with the acquisition of new information that increases the capacity for the movement of materials, people, and messages and continues until the enlarged capacity is fully utilized. These may be characterized as the adaptive, the growth, and the evolution propositions, respectively (1986: 7).

N. Levine elaborates on the relationship between culture and nature as interpreted by the human ecology perspective:

> Because culture determines the relationship between human populations and their environment, its study is of fundamental importance to the study of human ecology. All cultures change in time—sometimes rapidly, sometimes imperceptibly slowly. They change both in response to changes in the habitat and to imbalances within the cultural institutions that maintain the relationship to the habitat. For a culture is like an organism in one important respect: all elements of it seem to be seeking balance and integration with respect to one another. A particular feature of a culture or an organism can be judged only in its relationship to the whole and to the environment... (1975: 212–213).

The biophysical attribute of energy exchanges (energetics) is central to the human ecology model of social/ecological systems.[2] The relationship between energy flows and population increase (and by extension to development) is a major concern. Directly tied to the economic and ecological production and distribution of energy and other material staples in social systems is human population, which is "defined as an aggregate of organisms that have in common certain distinctive means for maintaining a set of material relations with the other components of the ecosystem in which they are included" (Rappaport 1984: 6). Population growth is a major concern for human ecologists and ecological anthropologists such as C. Geertz (1963) and E. Boserup (1965). The latter would rescue humanity from a Malthusian predicament by suggesting that increasingly intensive cultivation and the incorporation of marginal lands into the productive system in response to population growth. Geertz deals with the population problem through the concept of agricultural involution,[3] whereby,

> ...those cultural patterns which, after having reached what would seem to be a definitive form, nonetheless fail either to stabilize or transform themselves into a new pattern but rather continue to develop by becoming internally more complicated... all in an effort to provide everyone with some niche, however small, in the over-all system... (1963: 81–82).

In contrast, P.R. Ehrlich and A.H. Ehrlich (1973, 1991) suggest that the population issue can not be solved through increasingly baroque social production or distribution of material resources. Their concept of a "carrying capacity"[4] has been highly influential in terms of the environmental limits of growth. While the Ehrlichs' concerns are important in themselves, they are introduced here to highlight a distinction between the biologically-oriented human ecologists and other views along the symbiotic spectrum of the people/land relationship. Unlike cultural ecologists, the behavioral school, and political ecologists, human ecologists stress the importance of limits on material resources rather than social relations negotiating these resources as the key link between nature and society.

Ecological Anthropology

There is much overlap between the human ecologists and ecological anthropologists. As defined in behavioral terms by M.A. Jochim, ecological anthropology is based on the view that

> [H]umans are animals. They do interact with the natural environment, and their behavior does seem to show some patterning in relation to patterns in the natural

> environment. The ecological approach in anthropology tries to widen the viewpoint of the discipline to include (and by contrast, often to emphasize) this human interaction with the environment. Cultural behavior is studied not so much for what it is (learned patterns of behavior) as for what it does (provides one means of adaptation) (1981: 3).

The key point in this paradigm is the behavioral adaptation of human populations to their physical setting. Culture is ecological adaptation, i.e. human behaviors are adaptations to the ecosystems of which they are a part.

The stress is on self-contained entities, ecosystems, which are closed, internally regulative systems, i.e., cybernetic and tending toward equilibrium (Rappaport 1984). As stated by C. Geertz,

> The ecological approach attempts to achieve a more exact specification of the relations between selected human activities, biological transactions, and physical processes by including them within a single analytical system, an *ecosystem*....The concept of an ecosystem thus emphasized the material dependencies among the group of organisms which form a community and the relevant physical features of the setting in which they are found, and the scientific task becomes one of investigating the internal dynamics of such systems and the ways in which they develop and change...(1963: 3).

Geertz writes of the affinity between the ecological anthropology he articulates and the cultural ecology pioneered by J.H. Steward in the post-World War II era. Steward's work was indeed critical to the formulation of the cultural ecology school. But there is an even earlier antecedent to cultural ecology found in the work of a geographer, C.D. Forde. The rediscovery of cultural ecology by anthropologist Steward two decades after Forde's formulation exemplifies the disciplinary fracture between geography and anthropology. Independently scholars representing the two disciplines arrived at the same focus, cultural ecology.

Cultural Ecology

B.L. Turner II has defined cultural ecology as a "research perspective on nature-society relationships that are addressed largely, although not necessarily, at micro- and meso-spatial scales in non-Western settings" (Turner 1989). Turner, a geographer, cites the "strong links with anthropology, ecology, and resource specialties, not only in terms of common interests, but in general templates of problem solving" that exist with respect to cultural ecology. Regardless of disciplinary background, however, all cultural ecologists owe no small debt to the work of C.D. Forde (1934). The emphasis is on social adaptation to the physical conditions of a society, i.e., its habitat. According to Forde, the adaptive linkage between environment and society is cultural and economic. There can be multiple adaptive strategies within the same habitat, each with a different outcome. This notion has been referred to as "possibilism" which Geertz (1963: 1–2) defines as the view that the environment is "seen as not causative but as merely limiting or selective" in the determination of adaptive strategies.

There are no deterministic laws concerning the emergence of culture. There is "multilinear evolution" according to which societies inhabiting the same types of environmental habitats articulate different cultural relations with nature (Steward 1976: 18–19).[5]

Cultural ecology is not, however, blind to the evolution of the societies whose adaptive stability it emphasizes.[6]

The Ecosystem Approach

An important project related to, but critical of, the cultural ecology approach made its appearance in 1982 with the publication of R. Ellen's *Environment, Subsistence and System: The Ecology of Small-Scale Social Formations*. Ellen refers to his paradigm as the "ecosystem" approach, which he describes in this way:

> Theoretically, it [the ecosystem approach] has stressed the necessity for holism while focussing on specific relationships between human populations and features of their environment....It has stressed not only reciprocal causation but complex networks of mutual causality. It has focussed on system organization and properties: structure, equilibrium, change, the degree and forms of stability and the mechanisms which regulate the functioning of systems. It has emphasized the complexity of local environments and, technically, has focussed on the social significance of biological species. More generally, it has served to pave the way for the introduction and development of further systems concepts and ideas of energy flow and has revived an interest in Marxism and materialist dialectics (1982: 94).

The stress on "reciprocal causation" and "complex networks of mutual causality" marks the self-defined difference between the ecosystem approach and cultural ecology. They both share the search for culturally adaptive practices that preserve the ecosystem of which the human being, or social being, has an integral role. But Ellen also seeks to bridge the gap between cultural ecology and Marxism. According to Ellen, cultural ecology views the environment as determining the possibilities for social organization and culture. This contrasts with his position that the relationship between land and people is mutually formative. While the physical setting delimits the possibilities for human adaptation, new possibilities are created in turn through human agency:

> Thus the subtle interrelationships between natural and modified environments only serve to reaffirm that humanity and nature are not the two independent entities that are so often reflected in nature in ideologies; society is not the negation of nature. The problem is, as Marx pointed out, that Homo Sapiens is both part of nature, appropriating from it, and yet also capable of controlling it (ibid.: 15).

Ellen's ecosystem approach, rooted in the ecological anthropology tradition[7] attempts to transcend the latter by deriving an understanding of the people-land interface in a manner that is simultaneously informed by behavioral, systems, and Marxian approaches.

Regional Political Ecology and Political Ecology

Further along the spectrum of paradigms that characterize the relationship between society and environment as symbiotic is the political ecology school. One can differentiate it from the ecosystem approach in the emphasis the latter places on physical (e.g., energy) exchanges as opposed to the economic relations emphasized by political ecology.

Among the most distinguished of political ecology's proponents are P. Blaikie and H. Brookfield, whose *Land Degradation and Society* has been widely studied since its publication in 1987. The work represents one of the two major branches of the emerging political ecology school, which are differentiated from one another on the basis of commitment to Marxist orthodoxy. Blaikie and Brookfield discuss their more liberal, non-doctrinal approach to political economy in the following way:

> Our approach can be described as *regional political ecology*. The adjective "regional" is important because it is necessary to take account of environmental variability and the spatial variations in resilience and sensitivity of the land, as different demands are put on the land through time. The word "regional" also implies the incorporation of environmental considerations into theories of regional growth and decline.
>
> The phrase "political ecology" combines the concerns of ecology and broadly defined political economy. Together this encompasses the constantly shifting dialectic between society and land-based resources, and also within classes and groups with society itself" (1987: 17).

The political ecology approach endeavors to explain place-specific phenomenon in terms of global processes. In this way it is different from other paradigms previously described. For example, "...it is hypothesized that many areas of the Third World suffer from a set of related symptoms which combine the results of land degradation, political and economic peripheralization, stagnant production, outmigration and poverty. However, there are clearly important variations in the politico-economic and physical histories of peripheral areas" (ibid.: 18).[8]

M. Watts' represents the second branch of political ecology, which is explicitly Marxist and has spawned considerable debate. Perhaps most provocative and representative of Watts' thinking is the notion that "Properly defined, nature is internally differentiated and the subject matter of human ecology is accordingly *inner*actions with nature [emphasis his]" (1983a).

Accordingly, Watts distinguishes his viewpoint from the others on the spectrum of society-environment relationships on an ontological basis. He dismisses the separation of human beings from nature, ontologically as well as analytically. The society-nature relationship is not one of interactions between distinct entities but a symbiosis of interconnected parts.[9]

There are other differences between Watts' project and the previously discussed ecological approaches. The importance of labor as the vehicle for human inneractions in nature is complemented by the second concept Watts employs, intersubjectivity, which he attributes to A. Sayer and which further infuses political ecology with a "social" meaning. What Watts attempts is the creation of a unitary conception of human-nature relations that also incorporates the social essence of human existence—human life is conducted both *in* nature and *in* society.

Despite the different thrusts of their presentations Blaikie and Brookfield, Bassett and Watts share a common departure point for their analyses—social (i.e., socioeconomic) relations are critical to understanding the human-nature interface, and these relations are manifested at every scale, again in nested fashion, from the local to the global level. This is the hallmark of the political ecology approach.

Bookchin's Social Ecology

Much ink has been spilled in defense or condemnation of Murray Bookchin's seminal work on social ecology. A great deal of the fray has to do with Bookchin's provocative and militant style. Nonetheless, there are substantive issues that make Bookchin's approach a central one in discussions concerning sustainability.

Social ecology a là Bookchin is about the "belongingness" of society in nature. This belonging makes society's structures and institutions, especially those pertaining to social elites, exceptionally powerful in determining the welfare of the environment. The corollary is that nature, of course, has no less a stranglehold on society. The difference between the two is that society through its individuals, classes and elites has the capacity to exert conscious, that is deliberate, interventions in nature. There is simultaneously great creative potential in that difference as well as great danger as human beings can "open a remarkably expansive horizon for development of the natural world—a horizon marked by consciousness, reflection, and an unprecedented freedom of choice and capacity for conscious creativity" (Bookchin 1990: 37). On the one hand, they can dominate nature, a proclivity often shown and which "stems" from the "domination of human by human" (ibid.: 44).

In other words ecological problems are first and foremost social problems deriving from hierarchical relations in society. Environmental crises are mediated by the predations of elites on other classes and by their gluttonous usurpation of nature as property for their private enjoyment. Humanity can reclaim its humanity and enable nature to recover from its injuries only when freedom of choice is offered to all citizens.

Bookchin's approach, while strident and combative, contains a great deal of insight, of-this-world-ness and vision. It is, perhaps, the best known of a school of perspectives described by Merchant (1992) as "radical ecology" that includes deep ecology, spiritual ecology, green politics, and ecofeminism. Between these poles are the systems, behavioral and cultural ecology positions that stress equilibrium, adaptation, culture and values.

The above represents an attempt to define, a continuum of views on the land-society interface that perceives nature and people as mutually effecting. At one end of the spectrum are those paradigms—human ecology, ecological anthropology—which stress the ecological side of the balance. At the other end is R. Ellen's ecosystem paradigm, political ecology and Bookchin's social ecology, which stress social relations in the land-society integral, albeit while emphasizing different nuances.

Ecocentric thought includes the Gaia hypothesis, Lovelock's (1979) proposition that all of nature is a living whole (Miller 1991). The hypothesis has drawn a great deal of attention more as a curiosity than as an applicable theory. The Gaian notion has struck some as a modern paganism and romanticism. Many ecologists of the different schools described find the Gaia hypothesis irrelevant or, like Bookchin, repugnant and anti-social (Merchant 1992; Young 1990).

Parallel to the paradigms along the continuum of symbiotic approaches is a group of conceptual constructs that can be referred to as the ethnoscience/indigenous knowledge/eco-development school. These perspectives differ from one another in terms of their disciplinary heritage. More significant than their differences, however, is their agreement regarding the intimacy that characterizes the land-people interface.

ETHNOSCIENCE, INDIGENOUS KNOWLEDGE, AND ECO-DEVELOPMENT

In the 1985 re-issuance of her 1978 ethnography of an Andean peasant village, B.J. Isbell describes the contrary pressures operating on a community that is forced to Modernize.

According to Isbell, while some Chuschinos have conformed to the new wave, many, perhaps most, have deliberately resisted Modernization by having "increasingly closed in on themselves socially, economically, and symbolically in order to strengthen their defense against the encroachments of the outside world. Their overriding concern is to preserve their autonomy" (1985: 217).

Closure is but one of the strategies selected by traditional communities in resisting efforts to make them "ethnic internal colonies who have gained mobility into the dominant culture," (ibid.: 20) and whom "have been pushed continually into more marginal and less productive regions" and, at least in the case of Andean peasants, have become "the marginal agricultural majority...[who are] considered as part of a larger social unit, such as the nation, that has impinged upon its agricultural producers" (ibid.: 31–32).

Andean communities are not alone among indigenous peoples who face threats to the traditional society-nature bond linking them to their land. Some of these peoples, Amerindians, for example, have been decimated and their collective existence threatened by force and persecution emanating from expansionist conquerors. Others, like the Kurds, have achieved limited success in their strategy of armed resistance to assimilationist forces. The forces of Modernization, even when they are well-intentioned, impose a kind of violence that shears the society or a sub-group within it from its environmental heritage.

Many indigenous peoples have exercised resistance, both armed and non-violent, against centralizing powers threatening their life-place. Resistance in the face of perceived threats to ethnic minorities in the Philippines is described by J.N. Anderson in this way:

> The general attitude of the dominant lowlanders toward these upland minority peoples is essentially one of benign neglect; they harbor strong negative stereotypes toward them and consider them fair game in all economic and political dealings. Rights of the majority population for hydroelectric or geothermal power, irrigation water, tree plantations, and the like have routinely taken precedence. The destruction of habitats that support the ways of life of upland minorities has required them to make drastic changes just to survive.
>
> In response to ultimate threats to their resources and to their survival as peoples, some minorities have organized resistance movements....In Mindanao especially, but also in Northern Luzon, Samar, and Bicol, the Armed Forces of the Philippines' response to cultural minorities with force has spawned much wider conflict that had existed prior to martial law. This widening war itself has brought further ecological destruction to the uplands, especially in forested areas where rebels take refuge.
>
> The lack of protection of land rights has hurt Muslim and tribal populations alike. For all of the cultural minorities the loss of ancestral land presages the loss of ancestral culture (1987: 258–59).

Defiance by indigenous peoples has taken place violently around the world. An obvious question is: Why do such communities persist in asserting their collective existence, even at the risk of obliteration? Apparently "primordial ties" do exist among many peoples, including those in diaspora.

INTIMACIES OF SOCIETY-LAND UNITY

R. Riddell accounts for the tenacity of people-land relations by stating that "Marx observed that 'nature fixed in isolation from man is nothing for man,' a truism with a corollary, namely that people are also a product of their environment" (Riddell 1981: 53). This observation did not await Marx; however, it is as old as human literate self-reflection. W.G. Plaut, B.J. Bamberger and W.W. Hallo comment on the ceaseless Jewish commitment to the land of Israel as central to Jewish group identity across time and space:

> In the course of centuries, and especially in modern times, many Jews came to feel that God's role no longer needed to be considered in their relationship to the land. They were satisfied that history had forged an indissoluble bond between land and people and that as homeland and as the cultural and political center of Jewry it remained the focus of the age-old dreams (1981: 100).

Numerous Halachic (Jewish Law) dictates exist with respect to ecological practices concerning the land of Israel such as laying the land fallow for replenishment during the sabbatical year and injunctions against deforestation of fruit trees (ibid.: 1479). Scores of the 613 commandments that Jews are enjoined to fulfill can only be implemented in the land of Israel, that territorial context to which virtually all Jewish rite and lore refers. Jews, then, consider themselves incontrovertibly tied to the Land of Israel, expressing this connection during daily prayers and other religo-cultural practices.[10] Similar beliefs and practices of Native Americans regarding the emotive aspects of their relationship to ancestral lands is a subject much described in contemporary culture. Lands that are considered sacred and ancestral infuse meaning to human existence among traditional communities. Numerous examples can be cited.

A phenomena that has come to haunt the Modernization project is its belittlement of traditional knowledge systems, also known as indigenous knowledge or ethnoscience, along with native culture, generally.[11] As S. Amin states:

> A social reality exists when individuals are conscious of it and desire to express it; no right has higher value than such expression. Scientific analysis may provide

> an understanding of the objective conditions that create this reality, but it does not justify giving "prior warrant" to its expression. It is not the duty of thinkers and researchers (any more that of the authorities) to decree whether a reality (ethnic or otherwise) exists or not. That right belongs only to the people and to them alone...(1990: 95).[12]

Amin's statement reflects the right of peoples to exist as they choose. Modernization, however, has denied indigenous peoples these rights. The effects extend not only to the social group but to the entirety of their life-place, specifically their environment. Much of the shared decline derives from disruption in the application of indigenous knowledge systems. R. Chambers writes in this regard that:

> Centralized urban and professional power, knowledge and values have flowed out over and often failed to recognize the knowledge of rural people themselves.... These have been variously described as people's science, ethnoscience, folk-ecology, and village science. The ethno prefix is widely used, as in ethno-ecology, ethno-soil science, ethno-agronomy, ethno-anatomy, ethno-taxonomy, ethnobotany, ethno-medicine, ethno-linguistics and ethno-aesthetics. Others have written about indigenous technical knowledge which can be contrasted with modern scientific knowledge. More simply "local knowledge" has also been used (1983: 82).

Indigenous knowledge systems[13] preserve the equilibrium of life-place by using traditional values, practices and rites to govern the relationship between environment and society. The conception of two separate spheres of the sacred and the profane as posited in western thinking is permeable among traditional peoples. One of Modernization's most inglorious attributes is the commitment to replacing traditional ones with "modern," global (i.e., capitalist) ones. Yet many traditional societies define human and environmental welfare as intertwined. Their resource management regimes provided social stability for long periods of time.

Ethnoscience and Ethnoecology

Since the early 1970s development specialists, resource managers and human settlement planners have increasingly scrutinized what has become known as ethnoscience. C.S. Fowler defines ethnoscience as "the sum total of a group's knowledge, conceptions and classifications of objects, activities and events of its social and material world..." (1977). Ethnoscience is itself a broader umbrella of sub-fields dealing with particular fields, such as ethnohistory, ethnobotany and ethnogeography (Chambers 1983). Collectively these constitute the more general framework of ethnoecology:

> ...ethnoecology [is] a distinctive approach to human ecology that concerns itself with native conceptions of their environment. Its method, drawn principally from the method of ethnoscience, attempts to demonstrate the systematic relationships between native terminological systems for the environment and those conceptualizations. (ibid.).

Just as different flora and fauna populations coexist in the same ecosystem by filling complimentary niches, distinct ethnic groups may coexist in the same territory by maintaining different production systems, value sets and cultures. B. Spooner develops this notion in discussing the nomads and other groups of Baluchistan:

> The significance of the nomads for the future development of Makran [the southwestern division of Baluchistan, Western Province, Pakistan] far outweighs their numbers or their economic contribution. They are the only people who are likely to use some 90 percent of the territory of Makran. Without them the greater part of the population would be marooned in isolated oases, which on their own do not have the resources to be economically independent, and with increasing dependence on outside subsidies would gradually lose population to more attractive opportunities outside the province. With the nomads, the Baluch population as a whole forms an interdependent

> social and cultural, as well as economic and political, network covering the whole of the area. As long as the nomads are there, the whole of the area continues to be inhabited by people who consider it to be their territory. If the nomads leave, the settled population will see itself simply as an economically disadvantaged appendage of the national economy. As long as they remain, the total population shares a conception of ethnic provincial autonomy (1987).

Striking similarities to the situation described by Spooner in Baluchistan can be found with respect to Sudanese nomads (F.N. Ibrahim 1987) and for Andean peoples (Isbell 1986; Murra 1976). D.A. Posey describes the importance of indigenous knowledge in Amazonia.[14]

Indigenous knowledge systems have proven to be more insightful in certain areas than modern science. As R. Ellen relates, ethnoecology "has been a major source of evidence for reconstructing environmental changes, human migration and history, and the spread and introduction of new species, especially domesticates. On these grounds alone it would require no further justification" (1982: 218).

Ethnoscience and Gender

D. Rocheleau reports on a project in rural Kenya undertaken by the International Council for Research in Agroforestry and Wageningen University that revealed a gender pattern to ethnoscience knowledge and dissemination, one which Modernization has severely impaired. Rocheleau also cites differences in rights and responsibilities between men and women and the "spatial division of labor between men and women into rural and urban domains" as underlying the impact of gender inequities on ethnoscience, especially given the changes that have occurred in postcolonial societies. She concludes that:

> The mere recognition and documentation of survival as a gendered science in harsh and unpredictable environments (political, economic, ecological) may effect change at local and national levels. At best it may even serve to re-establish the legitimacy and strengthen the dynamism and innovative capacity of rural women's *and* men's separate, shared and interlocking knowledge as tools to shape their own futures (ibid.).

Others have commented that gender-patterned "development" distorts the traditional and social management of ecosystems, especially with respect to production and human reproduction[15] (Kelkar and Nathan 1991; Merchant 1989; Shiva 1989).

A cardinal lesson to be learned from ethnoscience and indigenous knowledge is that the human habitat is deeply in need of the people who have been an integral part of it. D.A. Messerschmidt amplifies this point in writing:

> Cultural experience encompasses a vast encyclopedic range of rules, customs, expectations, and things that tend simultaneously to order and reflect the ways in which we manage ourselves in societies. It also conditions how we manage, for better or for worse, the natural resources on which we depend for life and sustenance. And, just as biologists and geneticists are concerned with the rapid loss of genetic resources world wide, so too are social scientists concerned with the potential loss of sociocultural resources as a tragedy to be avoided at all costs (1987).

The reduction of cultural diversity, just as the loss of species and biodiversity are twin products of the replacement of ecosphere by technosphere. Mutations, the resilience of pests and infectious bacteria and viruses to pesticides and antibiotics, and birth defects clustered around areas of industrial toxins are examples of the transmorgification of organisms resulting from Modernization. Similarly, the proliferation of inner-city gangs, terrorism and suicide are indicative of the social costs of Modernization.

Indigenous knowledge and ethnoscience derive from empirical observations conducted over the course of generations concerning the physical habitat of a social formation. They are bodies of cumulative knowledge that reflect the relationship of a society to its terrestrial context (Klee 1980: 283). That relationship is continuous, not static, and changes over time as a function of modifications in ecological conditions, power relations internal and external to the group, and changes in technology. When either society or its habitat are disrupted, so is the relationship between them. Under such circumstances indigenous knowledge works fitfully if at all. Social instability ensues.

Modern science and engineering are examples of privileged twentieth century knowledge bases. Their prestige derives from the pivotal role they have played in the Modernization project. On the other hand, modern science has produced innovative technologies that greatly enhance human life. We have the opportunity today to choose what technological innovations are, on balance, sustainable (J.N. Anderson 1987; Bartelmus 1986; Bookchin 1990; Ehrlich and Ehrlich 1991; Murton 1980). So too, in those areas where life-place disruption can be reversed indigenous knowledge along with selected technologies and scientific information will prove essential to sustainable, endogenous recovery.

THE INTEGRITY OF LIFE-PLACE

The thesis advanced in this book is that place-specific solutions to global dysfunction are indispensable. Underlying this argument is the belief that society and nature are two parts of the same integral whole (Forde 1934: 466–67). The notion of the congruence between cultures and their physical setting is described by J.H. Steward:

> The culture area is a construct of behavioral uniformities which occur within an area of environmental uniformities. It is assumed that cultural and natural areas are

> generally coterminous because the culture represents an adjustment to the particular environment. It is assumed further, however, that various different patterns may exist in any natural area and that unlike cultures may exist in similar environments (1976: 35).

The culture area concept is not unproblematic. R. Ellen, for example, notes that the origins of the culture area concept began with the environmental determinist Ratzel (Ellen 1982: 8–9) whose geography has often been labeled racialist. Nonetheless, Ellen does not believe that this association discredits the concept's validity:

> There is no need to reject *a priori* the ideas of areas in which environmental and behavioral variables combine to give distinctive patterns, or reject the use of such features for the purpose of classification. This is both legitimate and necessary. It is important, though, to recognize the limitations of the term "culture area" as an ordering device and the arbitrariness of the classificatory factors employed, and to avoid transforming a category into an integrated system," (ibid.: 10–11).

While there are numerous difficulties in describing cultural areas—not the least of which involves the imprecision of boundaries and the rapidity with which societies have been Modernized (thereby altering both cultural and environmental characteristics)—there remains a continuity in change even after Modernization. In fact, the insistence on the retention of traditional ways of life is at least as strong as the desire of traditional peoples—who remain the majority of the world's population, particularly in the developing world—as the acquisition of modern amenities.

T.G. Verhelst writes of culture and development as follows:

> *[I]t is the idea of culture that gives both meaning and direction to economic activity, political decisions, community life, social conflict, technology, and so on. It is in fact culture that gives development its raison d'étre and its goal* [emphasis his].

> Culture...is, properly speaking, the basis of "development" (1990: 159–160).

But the recovery of traditional culture does not negate modernity, only Modernization.

> It is not a question of rejecting modernity *a priori* nor of returning to the past, accrediting it with unconditional value just as one had once done with Western-style progress. Rather, what is necessary is that one should become aware of the disasters incurred and of their causes. What is necessary is that the cultural communities who need it work towards new, *sui generis* conceptions of modernity. It is they who must reconstruct their societies, who must reconcile past and present [emphasis his] (ibid.: 62).

Verhelst's indictment of development as culturally imperialist is now accepted by many if not most development specialists. The socio-spatial context in which Modernization has been applied not only gerrymanders resource regions and ecosystems, but it creates new polities with very little social cohesion.

Endogenous recovery will take place only within the basis of society-environment integrity. This land-people relationship is encapsulated by the notion of Endogenous Recovery Region (ERRGNS) described below. Prior to examining the recovery process, a review of the evolution in consciousness found in the development literature is necessary.

NOTES

1. Unlike Alvin Toffler (1980), I view the technospere described in his book *The Third Wave* as an integral part of the problem of global dysfunction.

2. According to Hardesty,

> Recognition of a close relationship between energetics and human society goes back at least to Karl Marx....The ecological approach in anthropology has traditionally accepted such a "materialistic" interpretation of social and cultural behavior. Actually, it is more correct to see the ecological correlates of social organization as not just economic but as all environmental problems (1977: 75).

3. Gertz credits the American anthropologist Alexander Goldenweiser with introducing the concept of agricultural involution (1963: 81–82).

4. Erlich writes in this regard that

> ...the maximum size the human population can attain is determined by the physical capacity of the Earth to support people. This capacity...is determined by such diverse factors as land area, availability of mineral resources and water, potential for food production, and ability of biological systems to absorb civilization's wastes without breakdowns that deprive mankind of essential services....Whatever the maximum sustainable population may be, however, few thoughtful people will argue that the maximum and the optimum are the same. The maximum implies a bare level of subsistence for all. Unless sheer quantity of human beings is seen as the ultimate good, this situation cannot be considered optimal (1973: 226–227).

5. For further background on the methodology of cultural ecology, see K. Butzer (1989: 199).

6. For further background on methodology, see K. Butzer (1989: 199).

7. Ellen's paradigm also shares with ecological anthropology a common emphasis on energetics (1982: 130).

8. Hitherto employed in only a limited number of studies, the political ecology school has nonetheless elicited considerable interest. One element of its attraction is its nesting of detailed case studies within higher scale phenomena occurring at the meso-, macro-, state and global levels. An excellent, concise presentation of the approach is found in Bassett (1988).

9. The differences are described by Watts more specifically in this way:

   > In contradistinction to human ecology, which has tended either to humanize nature or naturalize man, a materialist perspective on society and nature is dialectical and internally related....[B]y acting on the external world and changing it, man changes his own nature. At the same time, the transformation of nature can only work with its given materials; human practice cannot transcend the laws of ecology, only the form in which these laws express themselves....(1983a).

10. For further discussions on Jewish ecological perspectives, see Daniel B. Fink (1993) and S. Gabbay (1994).

11. Belief systems provide the meaning and symbolism of indigenous knowledge. J.L. Beyer writes of Africa in this regard, that "Because economic activities are so intimately linked with belief systems and forms of social interactions, there are resilient elements in all societies which allow for continuity in the face of all kinds of pressure to change" (Beyer 1980). Similarly, B.J. Murton (1980) emphasizes the role of Hindu cosmology in South Asian ethnoscientific systems.

12. Unfortunately Amin applies these principles selectively, for example, belittling the non-Arab peoples—most notably Jews and Kurds—in the Middle East.

13. Chambers defines the informational content of what he calls "rural peoples' knowledge" in these terms: "Knowledge refers to the whole system of knowledge, including concepts, beliefs and perceptions, the stock of knowledge, and the processes whereby it is acquired, augmented, stored, and transmitted" (1983: 82–83).

14. D.A. Posey describes the difficulties of land management in a rain forest characterized by vast ecodiversity. His article is a reconstruction of how the indigenous system of land management in Amazonia offers sophisticated insight into field conditions:

    Fundamental to indigenous management is the reliance upon a wide range of plant and animal resources integrated into long-term exploitation of secondary forest areas and specially created concentrations of resources near areas of need (forest fields, forest openings, rock outcroppings, old fields, trailsides, agricultural plots, and hill gardens). Forest patches created by the Indians in cerrado/campo also provide dense concentrations of useful species. Maintenance, or more usually increase, in biological diversity is the key to successful indigenous conservation and exploitation.

    ...[I]ndigenous knowledge can help generate alternative philosophies for a more rational system of resource management in the humid tropics. The Kapayo are only one of many small enclaves of native peoples located in remote parts of the world, but the lessons they have learned through millennia of accumulated experience and survival are invaluable to a modern world in much need of rediscovering its ecological and humanistic roots (1985).

15. As Merchant writes,

    Production and reproduction interact dialectically. When reproductive patterns are altered, as in population growth or changes in property inheritance, production is affected. Conversely, when production changes, as in the addition or depletion of resources or in technological innovation, social reproduction and biological reproduction are altered. A dramatic change at the level of either reproduction or production can alter the dynamic between them, resulting in a major transformation of the social whole (1989: 18).

CHAPTER 4

THE ENDOGENOUS RECOVERY REGION

Recognition that the post-colonial state is inadequate to the task of development and that it is unable to provide for the self-determination of Third and Fourth World peoples (and to others elsewhere) necessitates an alternative framework to fulfill these requirements. This chapter is aimed at presenting such an alternative—the Endogenous Recovery Region (ERRGN).

The ERRGNs concept presupposes that *place-specific social, economic and ecological conditions are interrelated to and contingent on one another* and that *they collectively define an integral entity. The Endogenous Recovery Region is defined as that spatial area in which socioeconomic, cultural and ecological conditions converge and comprise a composite whole.* The degree of internal cogency of this composite will vary as a function of the degree to which the colonial experience and post-colonial Modernization has disrupted life-place. The social impact on a territory by its indigenous communities can be shown to persist even after considerable modification of local conditions have taken place.[1]

Essential to collective recovery is awareness of the Modernizationist roots that led to dysfunction. The antecedents of dysfunction due to the colonial experience have been a focus of study for some time, even prior to the independence of post-colonial states. Modernization has been less a focus of critique. However, as previously articulated, the shattering of endogeniety under colonialism, hypergrowth and life-place disruption are the products of Modernizationist ideology.

The economic premises of Modernization as critiqued by Friedmann and Forest (1988) include:

1. *Atomism*: Given the marginalist assumptions of neo-classical economics, conventional regional planning is concerned with units of capital (companies and firms) and labor rather than with societies and individuals;

2. *Statism*: The perspective of the conventional planner is that of the state center. Consequently, the regional population is reduced to an "object of exogenous actions by capital and by the state." It is an explicitly top-down approach, and;

3. *Nationalism*: Conventionally, "the national economy [i]s the largest relevant system for analysis." The regions are seen as moving toward national integration and spatial equilibrium. It completely overlooks place-specific conditions and or aspirations. Therefore, the state practices "politics of place" that subjugates the heterogenous life-places incorporated in the post-colonial state. Place is arrogated by national space in the interests of "progress" and modern "nation-building."

These phenomena create dualistic relations between center and periphery within the post-colonial society. They are also an integral element in the economic hierarchy of societies implicit to capitalist expansion and integration into one global system.

Center-periphery dualism (spatial disequilibrium) exists on all spatial levels—world, continental, national, regional and rural/urban. At all scales resource "leakage" fuels uneven relationships between marginalized areas and stronger state and international economic centers. Endogenous control over resources fades as a result, and the peripheralized area becomes increasingly vulnerable to economic and cultural intrusions. Herein lies the connection between dysfunctional development and social unrest since "dialogue" between center and periphery takes the form of a closed cycle of repression and rebellion.

Conventional development planning is premised on the cultivation of "growth poles" that eventually produce propulsive economic activity and the spread of economic growth.[2] For its first decades, development was synonymous with the modelling of economic growth:

> The technical literature and technically based policy in regional development...is positivistic and goal oriented, and rejects the qualitative as unmeasurable. It stresses multipliers, marginal and comparative advantages, and the like. This style of analysis has been greatly aided in recent decades by the joining of theory with mathematical formalism, and the availability of quantitative economic data together with extraordinary developments in statistical technique and computational power (Alonso 1988).

Growth poles refer to existing or future urban areas where industrialization would become progressively sophisticated and make ample use of regional resources through backward and forward linkages. Backward linkages implied the absorption of inputs produced locally; the production of output that could be passed on for further processing, marketing and sales and are the system's forward linkages.

Given that a major aspect of life in the post-colonial countries was its rurality, the "absorption" of inputs by industry translated into resource leakage from the hinterland to the center. Not only did growth poles fail to emerge in the periphery but economic energy was increasingly concentrated in the urban centers, especially in the primate cities.

THE RURALITY OF DEVELOPMENT

With growing recognition that the mass of people inhabiting the countries of the South —and the poorest people in these countries—were in rural areas, a review of Modernization and its industrialization bias began.

Eventually the changes in attitudes were profound relative to previous development theory: Agriculture was not to be regarded as a "backward" sector relative to industry. Further, the problems of the impoverished countries were seen as *structurally* distinct from the problems of agriculture in the countries of the North; for example, rising population dependency ratios meant that an ever increasing number of dependents required greater agricultural productivity per laborer on land that was declining in fertility. The major problems facing rural populations were not only their impoverishment but that of their land as well (Bunting 1975).

In response, the World Bank assumed a new approach to development in the poor countries and instead of promoting industrialization through urbanization, began stressing rural development:

> The objectives of rural development...extend beyond any particular sector. They encompass improved productivity, increased employment and thus higher incomes for target groups, as well as minimum acceptable levels of food, shelter, education and health. A national program of rural development should include a mix of activities, including projects to raise agricultural output, create new employment, improve health and education, expand communications and improve housing (World Bank 1975: 3–4).

The rural development approach adopted by the major NGOs during the sixties and seventies entailed an entirely new view of development focus and organization. It was directed toward rural areas where those in the poorest strata were to be assisted first. Further, planning was to be explicitly regional, focusing on rural areas. Quality of life, equity, local participation and empowerment were seen as playing no less important a role in social development than the achievement of economic growth. The movement toward such holism is perhaps best articulated by M. Cerena:

> The basic tenet...is that people are—and should be—the starting point, the center, and the end goal of each development intervention. In sociological terms, "putting people first" is more than an ideological appeal. It means making social organization the explicit concern of development policies and programs and constructing development projects around the mode of production, cultural patterns, needs and potential of the population in the project area (1985: Preface).

The evolution of an alternative rural regional approach to development is linked to increasingly comprehensive views concerning society. Among these, "bottom-up" (as opposed to "top-down") paradigms can be described, including the territorial, agropolitan, ecoregional and ecodevelopment perspectives. These approaches overlap extensively, and each contributes to the Endogenous Recovery Regional Concept under formulation here. It is useful to explore the basic assumptions they share, i.e., that development is best viewed as emanating from rural peoples themselves.

BOTTOM-UP REGIONAL DEVELOPMENT

Stöhr (1981) describes the bottom-up school as being oriented toward "the full development of a region's natural resources and human skills" (1981: 43). This is necessary to satisfy the basic needs of all strata in the regional economy and to reinvest surpluses regionally for the purposes of internal economic diversification.

Stöhr argues that social and ecological factors on the micro level are as good if not better forces for allocation than the elusive market force. Development should be promoted from within rather than being exogenously stimulated, since socio-cultural affinity and group solidarity permit easier access to land, equitable group decision-making, adaptation of regionally-appropriate technology, and emphasis on local projects supplying basic needs. Stöhr further contends that price mechanisms should be adjusted by the state center to subsidize peripheral areas while the latter remain in transition. Economic activities that can exceed regional demand should be fostered and export should be facilitated by restructuring rural-urban transport systems and improving intra-regional (village-to-village) communication and transport. Preserving or encouraging egalitarian social structures and group identity are regarded as integral to development.

Territorial Development

C. Weaver's (1981) conception of "territorial development" is similar yet distinct from Stöhr's bottom-up approach. Both respond to the "top-down" or "command" models of development whereby developing communities are directed "from above" by authorities who view communities as objects to be manipulated rather than the subjects of the development enterprise. Such paradigms are not monolithic. For Weaver, "territorial development"

> ...simply refers to the use of an area's resources by its residents to meet their own needs. The main definitives of these needs are regional culture, political power, and economic resources. Territorial development can be compared and contrasted to the idea of functional development, i.e. the narrow exploitation of a region's potentials only because of the role these play in the larger international economy... (1981: 93).

The territorial strategy relies on (a) job creation through the satisfaction of regional needs; (b) residentiary activities as the key to growth since the creation of value through local human labor and natural resources at the regional scale minimizes possibilities for leakages, backwash effects and domination; (c) the creation of regional infrastructure and community functions and non-dendritic transportation systems; (d) community education and territorial values; (e) decentralization in production activities and decision-making; (f) scales of activities that are small enough to be subject to local control; (g) natural resource use where the bulk of resources are used for meeting local needs; (h) the careful control of raw materials exports; and, (i) restraint and conservation in natural resource use.

Against this backdrop Weaver contends that the most important fulcrum for territorial development is collective will.

Agropolitan Development

The concepts introduced by Weaver in the territorial paradigm are echoed in the "agropolitan regional" approach articulated by J. Friedmann and M. Douglass (1981) and extrapolated by F.C. Lo and K. Salih (1981). Like territorial development, the agropolitan approach is regional-territorial in contradistinction to the spatial-functional emphasis of conventional regional science models.

Lo and Salih reaffirm the need for a regional approach, albeit an agropolitan one, arguing that[3] "existing solutions proposed for Third World countries are partial and exclusive, and that this is due to an improper articulation of the regional development problematique" (ibid.), Lo and Salih interpret the latter as involving polarization reversal, i.e., reversing imbalances between regions, that is, uneven development. They contend that studies of industrialization dispersion in Asian countries have failed to demonstrate anticipated spread effects from growth poles to impoverished areas.

Urbanization is seen as leading to the emergence of primate cities with the creation of two generally unarticulated sectors: a modern sector and an informal one. Further, rural areas are debilitated through outmigration to the economically dynamic cities. Concentrated, urban manufacturing involves no significant trickle-down effects to the rural poor, contrary to what Modernization theorists postulate (Hirschmann 1958). On this basis, Lo and Salih conclude that regional policy based on urbanization and industrialization is ill-conceived. The more efficacious approach to rural development employs rural-urban linkages in addition to the forward and backward production linkages of growth pole-dominated development.

The agropolitan regional strategy requires some degree of closure if the society is to be able to mobilize internally and benefit from its own resources, reduce its vulnerability to national and international economic market vagaries, and overcome dependency relations. Further, at the early stages of development economic linkages must be internal to the region rather than external to it. This applies both to production linkages and rural-urban linkages: "A necessary condition for the pursuit of rural development is...to reduce rural-urban distortions through the creation of rural-urban linkages on a symbiotic, equal basis at lower territorial scales of interaction...These rural-urban units may be called "agropolitan districts" (ibid.).

The essential features of agropolitan districts include: relatively small geographical scale; "a high degree of self-sufficiency and self-reliance in decision-making and planning, based on popular participation and cooperative action at local levels" (ibid.); diversification of rural employment to include both agricultural and non-agricultural activities, particularly small-scale industries; "urban-rural industrial functions and their linkages to local resources and economic structures," and; "utilization and evaluation of local resources and technologies" (ibid.).

The territorial dispersion of population in most poor societies should be geared toward rural areas and enhancing the rural-urban linkages: "Of particular concern in turning towards a rural basis in development are the socioeconomic relations within the agriculture sector and the role of rural towns and small urban places in directing a process of planning from below" (ibid.: 141).

The key to preventing regional "leakage" of resources is the achievement of polarization reversal through selective regional closure. Lo and Salih write that closure engenders the temporary economic protection of local regional industries through such mechanisms as:

1. incentives (credit, marketing aid, subsidies) for choosing indigenous technology based on local resources;

2. agricultural development through land reform and the provision of agricultural incentives such as price support, restructured land taxes, credit, collective distribution;

3. non-governmental, independent, peasant-based organizations that promote cooperative/collective participation in the management of development and institutional access to income generating assets;

4. the provision of basic needs and urban public services;

5. the devolution of state-bureaucratic functions to increase participatory democracy and local-level decision-making;

6. the arrest of capital flight induced by external forces, and;

7. the prevention of labor specialization, enabling fuller use of labor resources.

The Politics of Place

Even when broadly outlined as above, the agropolitan regional program obviously has political implications. J. Friedmann, who was among the major precursors of spatial regional science, but later adopted an explicitly political economy paradigm, views political struggle between center and periphery as the essence of endogenous development and regional autonomy. This paradigm is less concerned with economic optimization than with the "politics of place." Friedmann's approach emphasizes regional identities and their socio-cultural diversity. He argues for the direct, unmediated articulation of regional economies with global ones. Further, he maintains that only through focused struggle with the state center will peripheralized rural regions gain control over their economic and social conditions.

As seen by J. Friedmann and Y. Forest (1989) regional planning entails a shift away from functional regions and toward the territorial region, i.e., traditional social habitats. The conventional emphases on state and corporate economic initiatives is challenged: progressive socioeconomic change will derive only from the actions of a politically mobilized regional population engaged in proactive, self-prescribed development activity rather than being the objects of exogenous planning. Movement from long-term spatial equilibrium to spatial conflict and struggle (resonance of another type of conflict and struggle—class conflict and struggle—is unavoidably brought to mind) becomes imperative. Perhaps most profoundly, the economic growth objective is replaced by the concept that development should meet regional needs as determined by its inhabitants, rather than fulfilling priorities commanded by the state center.

A major claim made by the "politics of place" perspective is that only those regions which become politically mobilized can resist market forces that attack their "social and cultural distinctiveness as a life space for their population" (Friedmann and Forest 1989). The relevant actors here are regionally based social movements which seek greater freedom of cultural expression, greater political autonomy for their "homeland" and the defense of their "life space" against the threats of encroachment by outside capital (ibid.). Consequently, the emergence of political forces that seek to restore the place identity of the region is imperative. This enables the region to fulfill the two needs of developing communities as identified in the beginning of this book: sustainable, endogenous recovery and self-determination.

While regionalism is often viewed by researchers as an impediment to state political unity, Modernization, and development, Friedmann and Forest view realities quite differently. National states may attempt to ignore, destroy, co-opt, or accommodate regional movements; regardless of outcome, however, such movements will persist as the expression of territorially-based identities (ibid.). Regional movements respond to the neglect of place-specificity that is an integral art of national regional planning. Such planning is unavoidably political, albeit inadvertently so, in that it has had to "capture" the economic advantage of its various regions to compete in the international marketplace. Growth-oriented planning neglects regions where labor is abundant though capital is not, thereby creating the familiar "regional" problem of uneven development.

The politics of place requires that regional planning reject most economic models since they rely on investment in growth poles or privileged centers that do not exercise spread effects. Conventionally, the planner's client has been the state, and his/her tool of first recourse has been capital. A new place-specific planning methodology would entail the following conditions:

1. The state would no longer be the principal actor; regional populations would engage in direct action;

2. Since regional territorial movements are engaged in political struggle, planning is a politically engaged process;

3. Planners should deal less with policy analysis and design and concentrate more on formulating, bargaining and negotiating solutions;

4. The politics of place would emphasize endogenous development, with a reliance on regional resources (though not to the exclusion of other resources) and the self-organization of the regional population, particularly in cooperatives.

Ecodevelopment

The ecodevelopment approach is coherent with but extends beyond the agropolitan paradigm. While the agropolitan approach stresses the social relations of a group as it relates to a sense of place and to other social groups, ecodevelopment emphasizes the society-nature relationship from the ecological perspective. Instead of the agropolitan district, the ecodevelopment paradigm promotes an "ecoregional" focus.

P. Bartelmus, an architect of this paradigm writes:

> Aware of the relevance of ecological factors in the development process, the new planning concept of ecodevelopment has been advocated, in particular by UNEP. The same organization also offered one of the explicit definitions of ecodevelopment, as "development at regional and local levels...consistent with the potentials of the area involved, with attention given to the adequate and rational use of the natural resources, and to applications of technological styles...and organizational forms that respect the natural ecosystems and local sociocultural patterns" [(UNEP 1975)].
>
> This definition suggests a new "ecoregional" approach to development planning in referring explicitly to both ecological and sociocultural habitat. It is in this sense of reference to a well-defined ecoregion that the concept of ecodevelopment is used here... (Bartelmus 1986: 45).

The ecoregional approach, like the territorial and agropolitan approaches, informs the social ecology paradigm under formulation here. The former asserts the importance of ecosystem integrity as a focus for development regardless of the political and administrative boundaries that have been coarsely drafted on maps. Social ecology incorporates the notion of ecological integrity but fuses it with the social cogency of the ethno-national group or groups inhabiting a particular habitat.

THE RECOVERY OF TERRITORIAL LIFE

The concept of territorial integrity including social, cultural, ecological, economic and political dimensions, is close to what J. Friedmann and C. Weaver refer to as "the recovery of territorial life" (1979: 186). They elaborate by stating that

> Territorially organized communities may be conceived as arising in the intersection of three abstract spaces, each with its own attributes and describing a different dimension of communal life:
>
> — a common cultural space, because the claim to a sufficiency of livelihood implies a moral judgement that will be made only if there exists a tradition of shared symbolic meanings;
>
> — a common political space, because the equalization of access to the bases of social power requires a set of political institutions, actors, and roles with respects (sic) to which precise criteria of access maybe defined;
>
> — a common economic space, because the articulation of policies for a sufficiency of livelihood requires a finite set of interdependent productive activities and known levels in the development of productive forces.
>
> Although cultural, political and economic spaces intersect, they do not, as a whole, completely overlap. To the extent that they do, however, they trace the natural habit of a "community of destiny." Such areas of overlap may be designated as the primordial units of territorial integration (1979: 196–197).

Friedmann and Weaver propose five principles to guide territorial development:

1. development should aim at diversifying the territorial economy;

2. development entails maximizing the use of endowed resources in a manner consistent with principles of conservation;

3. development should encourage the expansion of regional and interregional markets;

4. development should be based as much as possible on principles of self-finance;

5. development should promote social learning[4] (ibid.: 197–198).

The above principles are incorporated in the concept of the Endogenous Recovery Region as elaborated below.

It is only when we can view society and the ecology of territory as mutually defined and mutually modifying that we will be able to fully understand the human-land relationship and the imperatives of sustainable, endogenous recovery. Economics mediates human-land relationships through a set of adaptations between the society (particularly between the individuals and sub-groups that constitute that society, specifically classes) and between societies. Therefore, economic relations are the *intermediary* relations in society-environment interaction; the society-environment relationship is the *primary* relationship between human society and nature. All too often, the intermediate and primary relationships are confused and economic relations are disproportionately emphasized.

Progressive socioeconomic changes that influence the economies nested across different spatial scales and integrated in the global economy must be focused on the social ecology of a given territory. Historically, when Modernization paradigms or those approaches that stress global economic growth are given free reign, indigenous peoples and other peripheralized groups are deprived of their identities and control of their lands in the "interests of the state." These groups become anonymous objects, unknown and insignificant compared to global process itself. The crimes against them, which are ultimately "crimes against reality," are repeated at every spatial scale and become manifested in impeded development and ethnic turbulence. Hence the need to realign space, place and empowerment through ERRGNs.

The notion of the Endogenous Recovery Region as an analytical framework and focus for implementation incorporates spatial concepts articulated by R. Hartshorne and D. Whittlesey concerning ethnic territory (see Chapter 3). The concept is territorial (place-specific, endogenously self-defined) rather than functional (exogenously circumscribed using economic friction models or similar criteria). Additionally, C.D. Forde and J.H. Steward's concepts of the culture area are also incorporated since the ERRGN is defined both by ecological and anthropological conditions of territory.

Modernization has been so comprehensive in its reach that all contemporary economic systems are integrated into a global framework. That is, all economies articulate with other economies at various spatial scales, downward as well as upward through different linkages (Dicken 1992; Knox and Agnew 1989).

While we can speak of an integrated, unitary global economy today, the manner in which different ERRGNs articulate with the world economy varies significantly. Endogenous Recovery Regions (i.e., denigrated life-places) are the loci of human activity that collectively forge global process. Endogenous Recovery Regions are units of collective territorial production and consumption, and economic analysis must begin with and be directed toward them. They are also generally culturally and linguistically distinct.

SPATIAL PARAMETERS OF THE ENDOGENOUS RECOVERY REGION

As described by D. Whittlesey (1954), a region is from a spatial point of view "an uninterrupted area possessing some kind of homogeneity in its core, but lacking clearly defined limits." Its essential feature is that it is an "area of any size throughout which a harmonious relationship between phenomena exists." There are differences between regions that are "homogenous in terms of specific criteria" (the "topical approach") and "total regions"—the "compage"—which are "differentiated in terms of the entire content of human occupancy of area. Such a region is an association of interrelated natural and societal features chosen from a still more complex totality because they are believed to be relevant to geographic study" (ibid.).

ERRGNs are compages: "The compage is, by definition, something less than spatial totality; but it does include all of the features of the physical, biotic and societal environments that are functionally associated with man's occupancy of the earth" (ibid.). The concept of the Endogenous Recovery Region is its definition in terms of congruence of the anthropological sphere, the environmental sphere, and the unique social relations developed as adaptations to these spheres (e.g., traditional economic and cultural systems).

The region is designated as such on the basis of place-specific environmental characteristics (e.g., soil and water conditions, biodiversity, climate, morphology, and topography) and the ecological "footprint" left on them by the social group(s) inhabiting the area. In most cases environmental continuities and discontinuities define the area encompassed by the region territorially. They do so by defining the natural "possibilisms"[5] for human adaptation present in the local environment while also representing the specific social ecology, economy and culture characteristics of the region.

Questions of Scale

The spatial reach and population size encompassed by any particular ERRGN is necessarily variable since the region is unique by definition. A region can include an area the size of Kurdistan (520,000 km^2 with a population of approximately 25 million), or, conversely, micro-territories,[6] even on the district, sub-district or borough level. Because development difficulties emerge when economies of scale are not met, inter-regional economic cooperation may prove indispensable to socio-political and territorial decentralization (Ronen 1979). Cooperative, inter-regional resource sharing in areas too small to sustain independent economic systems is implicit to the concept of ERRGNs. Such solutions are especially appropriate in cases where neighboring societies make use of a common resource base, e.g., a river system, a forest, or a mineral deposit.[7]

Each Endogenous Recovery Region is socially distinct in terms of ethnic relations, class stratification, racial relations, gender patterns, culture and sub-cultures, creed and world-view: collectively these social dimensions constitute the social ecology of the region. Indigenous knowledge mirrors the geographical features of place, and converts them into behavioral, linguistic and cultural symbols that can be understood and communicated among a people and used to foster a sustainable relationship between society and environment. Values, culture, norms, mores, rites, religion, indigenous knowledge, lore, and language are manifestations of the indigenous social ecology of the region. These palpable expressions of social cohesion echo the environmental realities present in the region. For instance, the variety of deities worshipped by different societies generally reflect key aspects—rain, sun, mountain, beast—of the local ecology.

Acceptance of social differences and recognition that endogenous recovery based on group self-determination is imperative to both social and environmental welfare. As discussed later, this may very well involve the reconstitution of the politico-spatial entities known as nation-states in poor countries, since these polities diverge widely from inherent territorial realities. Such reconstitution, as recent events in the former Eastern Bloc show, are not as inconceivable as would appear at first glance. Quite the opposite—ecoregional movements are on the ascent and resist the centralizing forces that have eclipsed them in the name of Modernization.

NOTES

1. An example of this is found in Kurdistan. Both Iraq and Syria have "Arabized" formerly Kurdish areas by resettling Arab peasants in Kurdish areas and displacing the indigenous population. Turkey has also adopted the strategy of forced "deKurdification" in part through the Southeast Anatolia Project. The latter is a massive irrigation and hydroelectric generating scheme that will radically alter the landscape of this Kurdish

region and displace thousands if not tens of thousands of Kurdish peasants.

2. The writings of Perroux, Boudeville, Isard, and Hirschmann are representative of the growth pole approach that is largely based on Christaller's central place theory.

3. According to Lo and Salih, "Located in this proper and wider context, regional development—like economic growth, cultural development, ecological balance, and structural transformation—is an instrumentality, the necessary means to achieving the goals of human development" (1981).

4. "Social learning occurs whenever the institutions comprising the agropolis show an enhanced capacity for dealing with the problems that confront them" (Friedman and Weaver 1979: 202).

5. As described in Chapter 3, C. Geertz (1963: 1–2) defines as the view that the environment is "seen as not causative but as merely limiting or selective" in the determination of adaptive strategies. There are no deterministic laws concerning the emergence of culture. The social ecology approach defined here repudiates environmental determinism.

6. As in the Gaza Strip.

7. The Dead Sea area at the border of Jordan and Israel is now being jointly planned by specialists from both countries. Negotiations are proceeding to determine the equitable distribution of natural resources, such as water, given their scarcity in both countries.

CHAPTER 5

ARTICULATION OF THE SOCIAL ECOLOGY THESIS

In suggesting that the Endogenous Recovery Region is both the tangible expression of the society-nature integral and an efficacious framework for endogenous recovery, one wonders why the integrity of the land-people relationship is so critical. Why would congruence between ethnicity and territory influence progressive socioeconomic change?

Potential or partial explanations include the notion that ethnicity is defined by continuities of language and culture, the media of discourse necessary for social life (Deutsch 1966). Ethnicity can also be described as the expression of social continuity linking generations, providing existential meaning to the individual, and orienting the individual and social group to their anthropological and physical milieu (Enloe 1973: 15; Weitz 1986: 15–45). While western Modernism may belittle the importance of the existential or spiritual basis of group solidarity, such solidarity is certainly significant with respect to the traditional societies of the Third World. Tivey writes in this regard that

> Nationalistic products do not merely supply wants; nationalism characterizes a want in itself. It forms, names, embodies, an emotion that is widely and very deeply felt; a feeling that a person belongs to something, is not alone, is part of a great community. Sometimes this feeling is regarded as non-material and therefore non-economic. The inference could not be more foolish (1981: 69–70).

This psychological need for nationalism is similarly described by Gordon in his work on self-determination, where he writes that the "collective memory" of a people is the basis for its "identity, [which is] a reservoir upon which it can draw to give itself meaning, and a destiny..." (1971: 3).

Ethnic identity, then, appears to be at the very least a force capable of surviving concerted efforts to suppress it. This contention is supported by the deepening cracks in multinational, formerly communist states such as Bulgaria, Czechoslovakia and the former Soviet Union. The same holds true in Chinese-occupied Tibet, where subjugation has prompted normally pacifist Buddhist monks to lead the opposition against foreign domination. So, too, Fourth World nations embedded in post-colonial states from the African Sudan to Kurdistan, from Bengal to the hill peoples of Burma and Thailand, from Sumatra to southern Mindanao continue to rebel against the centers of the post-colonial state that claim them as citizens/subjects. Whether it is looked upon with favor or not, ethno-nationalism appears to be as deep a social phenomena as relations of productions or the political rivalry of ideologically opposed groups. This fact can no longer be denied. Such solidarity has both material and intangible (but nonetheless real) social and psychological manifestations.

THE MATERIAL MANIFESTATIONS OF ETHNO-NATIONALISM

There is growing recognition that ethnic solidarity reflects concrete, material aspects of the group's physical setting. Each society and its ecological setting appear to have a unique, inextricable relationship historically defined by the particulars of locale. This is not to say that peoples can be ordered in a typology based on the physical setting in which they live or any other criterion; such thinking is anachronistic and potentially racist, as found in the geopolitical thought of environmental determinists such as Montesquieu, Ritter, Ratzel and Kjellen, and Haushofer. Rather, the correspondence of anthropological factors with ecological ones expresses the symbiotic relationship between societies and their habitat. As conceived by R. Peet,

> Place-specific natural determinism is the fertile ground for geographical investigation. There are several dimensions to this place-specificity. First, the nature of nature varies between different regional, and even local, environments—if reflection on nature is a primary source of intellectual life, then reflection on a particular nature can be seen as a primary source of regional consciousness and art forms. Second, humans encounter not only local nature, but a wider natural environment, through many kinds of spatial interactions, from migration to the diffusion of cultural traits—hence the aspect of relative location is significant. Third, the relation with nature is always mediated by socioeconomic forces and institutions, of which the forces of production applied to natural resources are an extremely important component. As the level of productive forces available to any regional social group increases, determination by nature decreases. Space thus becomes a mosaic of different levels of determination, similar to the different levels of economic development (1986).

Similarly, the neo-Marxist mode of production school views social formations and their adaptation to habitat as a critical subject of study (Althusser 1971; Bennholdt-Thomsen 1982; Byres 1985; Roseberry 1988; Soiffer and Howe 1982). Adherents of this perspective engage in empirically detailed analysis of small-scale, generally peasant social formations. They study the economic articulation of local level communities within a nested hierarchy that is vertically structured by meso-, macro-, and global processes.

Neo-Marxists are not alone, however, in stressing the tangible manifestations of ethno-national groups. K. Deutsch views society as a "single social reality" constituted by linguistic, cultural, geographic and economic continuities. Societies are delimited by "relative discontinuities" in geographic, ethnic, resource, labor and market dimensions. "Units of human geography are bounded by relative discontinuities," which determine zones that "bound a country...." (1966: 41).

Likewise, development economist A.O. Hirschmann in elaborating on his concept of the "tunnel effect" in which development dynamism is produced by the relative socioeconomic mobility of different social strata writes that the "more or less unitary character of a country is probably the most important single criterion for appraising the likely strength and duration of the tunnel effect" (1981: 51).

Cross-disciplinary research conducted from a variety of theoretical perspectives has reached a similar conclusion: anthropological, economic and ecological factors are interdependent and should be considered, analytically, as different manifestations of one distinct unit. Ethnic identity reflects factors that inextricably link peoples and lands. This seems as much a fact of nature as the law of gravity or the attributes of a Gausian curve describing random distribution.

As previously discussed, the intimacy of the land-people relationship is expressed in bodies of knowledge variously known as indigenous knowledge, rural peoples' knowledge, and ethnoscience. Indigenous, albeit informal, knowledge among traditional peoples about the environment in which they live is, in many cases, superior to academic knowledge, or is at least a valuable adjunct to it (amounts of political autonomy (Chambers 1983: 75; Cooke 1989; Lea and Chaudhri 1983: 333; Posey 1985; Richards 1985; Rocheleau 1991).

Additionally, as M.E. Lopez (1987) and J.N. Anderson (1987) have written with respect to a Filipino case, that there is evidence of correlation between ecological degradation and the destruction of indigenous cultures, especially concerning land use. Similarly, B. Spooner (1987) describes how development, ecological factors, and the survival of indigenous peoples and lifestyles are inter-connected with respect to nomadic and sedentary communities in Baluchistan. In environments that have not been grossly transformed it appears that indigenous adaptive practices represent the optimal use of local resources. When other practices are applied which result in the destruction or subjugation of indigenous groups, environmental degradation inevitably occurs.

D.P. Lea and A.M. Chaudhri maintain that "Rural socioeconomic systems, and indeed belief systems, are an integral part of the management, use and control of land..." (1983: 333). In the post-colonial state, however, resource bases are also divided; unitary habitats such as deserts, woodlands and coastal zones have been disrupted by the borders of states wherein different policies are applied. Incompatible policies implemented by regimes governing different parts of the same ecosystem frequently produce conditions that are incapable of sustaining the populations that have been living in symbiosis with their environments for millennia. As a result, environmentally adaptive knowledge and technology is lost.

LESSONS OF THE SOCIETY-ENVIRONMENT RELATIONSHIP

A review of the literatures reveals that with few exceptions current theory attempts to explain the Third World problematic (i.e., environmental degradation, ethnic unrest and impeded development) in terms of state policies. Conventional perspectives emphasize the role of regime in post-colonial states and maintain that developing societies will advance if governments apply sound, Modernizing policies. Accordingly, the state is as good or as bad as the regime that governs it. However, the state cannot be defined solely on the basis of regime. Regimes and their policies are exercised over space; people, in turn, inhabit that space. Consequently, territory and the social characteristics associated with it constitute the "arena" in which regimes institute policies aimed at development.

During the past two decades reexamination of the inter-related spatial and social aspects of territoriality has taken place. Consequently, the implications of what B. Anderson called the "isomorphism between each [colonial] nationalism's territorial stretch and that of the previous imperial administrative unit" (B. Anderson 1983: 104-05) have received broader attention. These new approaches are refocusing theory with respect to the anthropological, ecological and economic patterns characteristics of a territory in attempting to explain the Third World problematic.

COLONIZED SPACE AND THE IMPEDIMENTS TO DEVELOPMENT

The recognition that ethnic groups have a social and territorial individuality that has developed historically and which continues to exist is a profound one: such identity appears to persist even where there has been considerable change in landscape. On the whole post-colonial states have not succeeded in subsuming the continuity and individuality of the ethnic components inherent to the heterogenous societies they encompass. This ethnic diversity is expressed by distinct anthropological patterns including language, culture, belief-systems, life-styles, life-cycles, gender patterns and class structures. These anthropological patterns are strongly associated with the territorial setting in which the ethnic group resides or with which it identifies.[1] This unique anthropological/territorial association is manifested by the socio-spatial patterns or structures of a given area; socio-spatial patterns, in turn, demarcate areas that are termed here Endogenous Recovery Region (e.g., Kurdistan, Baluchistan, Tibet, and Timor).

How the social group uses the ecosystem of which it is an integral part and how it applies the system of production in adapting to its surroundings define the individuality of the ethno-region. These ecological and economic adaptations are referred to as the social ecology characteristic of a Endogenous Recovery Region. Examples of such relations would include hunting and gathering undertaken by clans, seasonal nomadism and animal herding organized by tribe, sedentary agriculture in hamlets that pay tribute to a feudal nobility, and town mercantilism where relations of production are caste-bound. Unique social organization (e.g., institutions) and cultural content, on the one hand, and ecological and economic relations (e.g., class structures and gender patterns) on the other, are intrinsic parts of a unitary human/territorial whole.

Inattention to the existence of historical ecoregions and their distinct characteristics obscures underlying realities that are fundamental to sustainable social change. To define a territory spatially is a necessary but insufficient condition for endogenous recovery and self-determination; the spatial dimension of territory cannot be separated from the social and ecological dimensions. Territory cannot be treated abstractly; it is inseparable from the people residing thereon or from their unique economic and ecological adaptations (i.e., social ecology relations). Therefore, progressive change cannot be sustained when *socio-spatial consonance* (i.e., the cogency of a society and its physical setting) is absent, or when an organic relationship between the society and its environment is lacking.

Under imperialism, "development" of colonized space took place according to the interests of the metropolitan powers (as opposed to those of the indigenous peoples). Such "development" required the creation of colonial polities whose spatial and, by extension, social parameters were determined by the economic, geostrategic and ancillary administrative requirements of imperial pursuits. Specifically, colonial state structures were created to advance the efficient extraction of human and material resources from the colony. The socio-spatial organization of the colonies frequently divided ecoregions by previously unknown borders. Consequently, colonies were composed of ethno-regional fragments that were anthropologically divided and characterized by diverse sets of society-nature relations. For example, in the former British colony that became Kenya, there are Somalis and other Hamitics in the northeast, Massai in the south, Turkana in the Rift Valley, and Suk and Luo in the West. Their environments range from coastal zones to highlands, from desert to forest, with each group having a distinctive cluster of ecological and economic adaptations. Conversely, Somalis reside on both sides of the Kenyan/Somali border and Massai inhabit adjoining regions of what has become Kenya and Tanzania (i.e., state borders cut through Endogenous Recovery Regions). Economically, diverse traditional and market sectors coexist in an unarticulated fashion.

Post-colonial states are the institutionalization and indigenization of former colonial space: such states are on the whole socio-spatially conterminous with the former imperial polities (see Figure 5.1) Consequently, in many respects the post-colonial pattern of development follows the patterns, institutional as well as economic, set under imperialism. Indigenous elites concentrated in the state center (usually the capital city, which is often the primate city) formulate policies for adoption in a territory whose delimitation was established according to the imperial mode of development. In such cases, national unity is not a function of socio-spatial congruence but of a post-colonial legitimizing creed such as anti-imperialism, pro-Westernism, Marxist-Leninism, or Islamic fundamentalism. However, just as colonies were socially and ecologically fragmented, post-colonial societies and territories, too, are fragmented. Since post-colonial development is largely an extension of the imperialist mode of development, dualistic relations continue between the countries of the North and South, between the modern and traditional sectors within a developing society, and among ethnicities. Weak states are inevitable in this context.

Consequently, from the socio-spatial perspective post-colonial development is a continuation of the imperialist mode of development. Sustained development progress is unattainable since current conditions are continuous with exogenous ones (i.e., colonial origins and orientation). In this light socio-spatial restructuring of states is necessary to fully utilize the social ecology relationships that are central to Endogenous Recovery Regions but which have

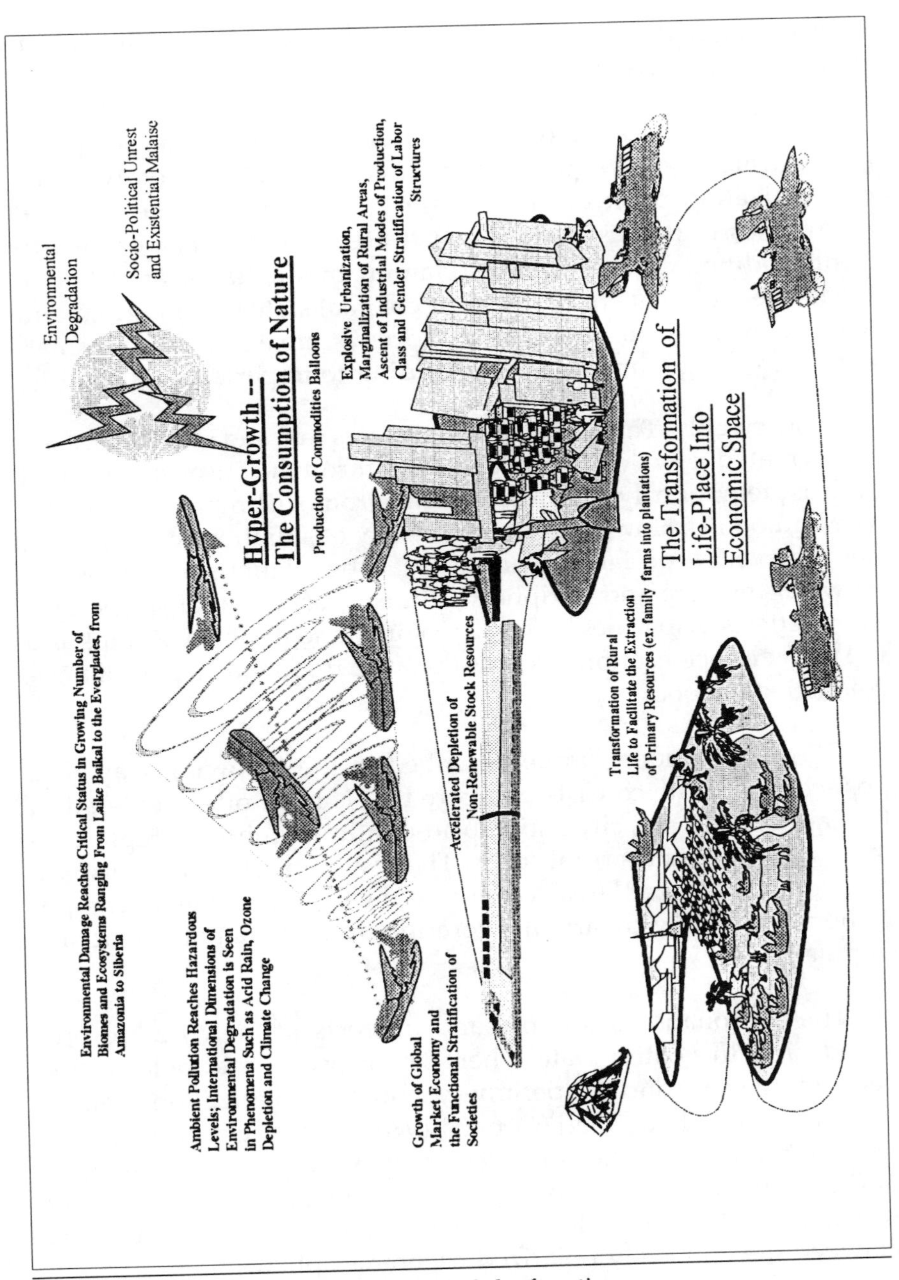

Figure 5.1 The development of global dysfunction.

been obfuscated by the post-colonial state structures into which they have been divided.

Another implication of the continuity between the socio-spatial configuration of colony and post-colonial state involves the decolonization process. The transfer of power from the imperial metropoles to domestic elites constituted a necessary but insufficient condition for recovery. By retaining the spatial composition of the former colonial entity, the post-colonial state has inherited and institutionalized the a-historical "society" created by amalgamating or dividing ethno-regions according to imperial need.

Conceivably, Third World nation-states can, and in certain cases, have created new "national" identities through consensus reached (at least temporarily[2]) between ethnic groups (Zimbabwe may be[3] a case in point). However, in the majority of situations consensus is not achieved, especially given the dualistic relations that transpire between the core and periphery of most post-colonial states. This often reflects disparities between majority and minority groups and the uneven socioeconomic status that is so frequently a part of post-colonial statehood.

These conditions can be described in the following way. The post-colonial state consists of a core that is most often centered in the capital (primate) city and a socio-political hinterland representing part of the state's rural space. The rest of the state's rural space and some urbanized areas constitute the state periphery. The periphery frequently dwarfs the core in terms of spatial extent and population.

The ethnonational identity and interests of the core are projected onto the entire state. There is socioeconomic polarization between the core and the periphery, and this polarization is ethnically or regionally defined (Copans 1980). Dualistic economic relations detrimental to the rural periphery are typical in such situations, again using elites of the subjugated groups (such as the Kurdish landlord class discussed later in this volume) as a bridge between the ethnically polarized economic spheres.

Over time, the stratification of life-places in ex-colonial states becomes untenable since labor, material and capital resources move from the periphery to the center and deplete the former; from the rural regional point of view, peripheral areas suffer a systematic and substantial "leakage" of resources to the state and, by extension, to international centers. To maintain these processes the ruling elite adopts draconian measures. Extreme economic and social stress either crushes the subjugated population or forces them to rebel. Under such circumstances the center—allegorically as well as literally—does not hold. Social disorder, economic duress, and political instability result and possibilities for development are impaired, if not precluded.

To summarize, the socio-spatial characteristics of a state and the geo-ethnic integrity (as opposed to political and administrative unity imposed from the center) of state territory are significant determinants of endogenous recovery and self-determination. Where the state is constituted by multiple ecoregions and/or fragmented regions,[4] the potential for recovery is seriously compromised since a myriad of anthropological, economic and ecological relations remain fragmented. Political unrest, ethno-regional stratification, and environmental degradation are inevitable factors of a destructive dialectic inherent to such conditions.

Shifts in development thinking have paralleled the evolving consciousness concerning the socio-spatial maladies that derived from the decolonization process. Development (and post-development) theorists have reconsidered the urban-biased, aggregate, state-level perspective implicit to the Modernization models and now stress the rural region qua life-place as the focus for development. In so doing they have forged a significant re-conceptualization of development's territorial context from the abstract, macro aggregate level to a more "human" scale (Seers 1983).

The redefinition of the socio-spatial focus of progressive change has led to a break with the Western-biased economic growth and Modernization models; the latter still informs much exogenous intervention, though the new approaches are gaining prominence. The notion that socioeconomic advancement must be achieved endogenously rather than being dependent on exogenous socio-economic manipulations is common to many of the new theories and represents a conceptual revolution in development thought (Weaver 1981). New concepts of "region" have emerged: the region is conceived of as that area where common background and mutual interest among the inhabitants, as well as ecological coherence and uniformities in the system of production, enables endogenous recovery to take place.

All inhabited territories are potential Endogenous Recovery Regions. In surveying the literature, however, it is clear that what has been termed here Endogenous Recovery Regions and the social ecology relationships (relations between the social group(s) and its/their environment) intrinsic to them are largely neglected as a factors in the development process. This is paralleled by authoritarian state structures that exclude the involvement of citizens or particular groups of inhabitants in the affairs of civil society.

THE SOCIAL ECOLOGY PROGRAM

Territorial planning is based on political economy principles as well as conceptual innovations offered by the ecodevelopment (Bartelmus 1986), agropolitan (Friedmann and Douglass 1981; Lo and Salih 1981) and territorial (Weaver 1981) paradigms. These perspectives stress endogenous, ecologically-sound development strategies spatially oriented toward regional and sub-regional scales (C.Y. Thomas 1974). One can synthesize from them principles for the evolution of endogenous recovery. Accordingly, the principles of the social ecology approach are listed below.

I. Neither environmental determinism, which attributes socio-cultural phenomena solely to the force of physical setting, nor extremist social constructivism, which seeks to explain landscape, technology, culture and development solely in terms of human agency are tenable descriptions of the nature-society relationship.

II. While the relative proportion by which nature and society mold the other is a matter of legitimate debate and has resulted in the diverse paradigms discussed above, the human-land relationship is an integral one, i.e., land and society are characterized by reciprocal causality. Accordingly,

   A. Social groups do adapt to their environment and such adaptation is manifested by traditional:

      1. Patterns of human settlement or nomadism;
      2. Economic systems (mode of production, distribution, and relations of production (class structure);
      3. Land and resource use;
      4. Social relations (e.g., gender relations; social stratification);
      5. Culture;
      6. Belief systems and values;
      7. Technology;
      8. Energy production and consumption;
      9. Nutritional practices;
      10. Health care practices;
      11. Civil institutions;
      12. Legal systems; and
      13. Indigenous knowledge.

   B. As societies adapt to their environments, they also alter the environment, which feeds back to change their adaptive practices.

   C. Change in one of the above aspects produces change in all other areas, i.e., traditional society represents a unitary social reality.

D. All of these dimensions are inter-connected and constitute a matrix defining areas of relative continuity.

E. Societal continuities remain amidst change, and exogenous change that threatens such continuities is frequently resisted. Ethno-nationalism may appear to be merely an ad hoc instrument used to mobilize oppressed peoples against oppressors; while this may be true of the assertion of political ethno-nationalism group identity, solidarity is a constant if latent aspect of individual identity; the conscious activation of such identity as a mobilizing creed arises in the face of threats by another group; in this case, ethno-nationalism is an activist manifestation of deeply-rooted latent identity.[5]

III. All societies change over time, although this is frequently imperceptible in long-standing cultures.

IV. The sine qua non of *traditional* societies is found in the relative stability of their adaptive practices.

A. Significantly, extroverted elites among the dominated group itself may be a key agency in the appropriation of surpluses and the transfer of resources from the peripheral group to the state center.

B. Internal class differentiation may take place as a result of the state center's encounter with the expanding capitalist world-system and the incorporation of extroverted mercantile or landowning elites within the subjugated group. This will result in changes in the mode(s) of production; the multiple modes of production in post-colonial states are usually unarticulated, and post-colonial economies are generally dualistic.

C. Subjugation of women in traditional societies as a link in the chain of oppression faced by the subjugated group.

D. Long-standing conflicts among ethnic groups may exist, but seldom do they lead to the extinction of any of the groups.

V. The sine qua non of *modern* societies is the relative rapidity of change in their adaptive practices.

A. Encounters with modernity, particularly among intelligentsia schooled in the western tradition, may effect change, especially internal change against exploitative feudal powers or the prosecution of ethno-nationalism against an oppressive, state center.

B. The agents of such change may be secularized intellectual leaders who challenge the traditional leadership of the group's ruling classes, or who may form alliances (for tactical or strategic reasons) with the traditional leadership in order to fight the state center. The traditional leadership, for their part, may agree to alliances with politicized groups with whom they have no ideological affinity for tactical reasons, e.g., since modern ethno-nationalists often constitute the most politically dynamic of the ethnic group's human resources, traditional leaders may find it useful to ally with them.

1. Allied ethno-nationalists may later find themselves exploited and/or suppressed by the traditional leadership; and

2. Alternately, upon assuming power modern ethno-nationalists may circumscribe the activities of traditional leaders.

C. Subjugated groups may often adopt left-wing ideological positions as a result of heightened consciousness among the intelligentsia concerning oppression; such ideological influences may filter upward to the traditional leadership with whom the emerging leaders are aligned for the purpose of fighting the state center.

1. Ideological and/or economic and/or tactical differences among ideologically-informed groups may result in political segmentation within the ethnonational movement—such conflicts may lead to fierce struggle.

2. For ideological, more banal reasons, or sometimes for lack of choice, modernized elements of subjugated groups may opt for assimilation or even alliance with the state center. The state center will often cite such assimilation to bolster claims that policies aimed at group pluralism have been adopted by the state.

3. An incipient indigenous bourgeoisie may often contest ethno-nationalism, particularly if it threatens economic relations with the state center. They may establish parties reflecting their class interests, and these parties may be either friendly or hostile to the liberation struggle of the oppressed group, depending on economic rather than ethnic interests; full-fledged class conflict may ensue or transient intra-ethnic conflict may result.

A. Inter-ethnic conflict may produce internal economic domination by one group over another or others.

B. Inter-ethnic conflict may create cultural and linguistic domination by one group over another or others.

C. Secular ethno-nationalists or other moderns often find their base of support in urban areas while traditional leaders will find their's in rural areas.

D. Traditional leaders may rely on religious or other traditional creeds and symbols to mobilize their populations.

VI. Disruption of traditional land-people relationships is generally accompanied by environmental degradation.

VII. The multilinear evolution of cultures does not preclude the cross-cultural transmission of traits, which occurs over a greater extent of space and with greater rapidity as a function of higher forms of technology.

VIII. All societies currently intereact economically with other societies. Modernization has produced patterns of economic exchange that have disrupted traditional society-environment relations and have established a hierarchy (in terms of benefits gained relative to resources contributed) of societies at all spatial scales, i.e., on the local, sub-statal, state and international levels. This has led to

A. Impoverishment and blocked development,

B. Environmental degradation,

C. Social and cultural crises (e.g., class, regional and ethnic polarization, gender subordination), and

D. The economic encapturement of previously autonomous regions within the realm of a state that is being drawn into the system of international economic expansion.

IX. Endogenous recovery must be self-induced and self-directed—external manipulation of societies and their environments often result in unbalanced exchanges to the benefit of the agents of change and to the detriment of the peoples and environments onto which change is imposed.

X. Large-scale environmental degradation and social advance should be regarded as mutually exclusive—the latter should not be attained at the expense of the former.

XI. Ethnoscience (indigenous knowledge) and modern science are enhanced when they are mutually informed and when both are applied concurrently.

SOCIAL ECOLOGY AS POLITICAL STRUGGLE

There are political consequences to these principles. The interests of progressive social and economic change may necessitate that the governance of Endogenous Recovery Regions rest with indigenous territorial authorities empowered with varying amounts of political autonomy. This in turn may require the reconfiguration of post-colonial state borders or create new boundaries within existing states. The implications of such scenarios are weighty as they threaten the continued existence of post-colonial states. Nonetheless, one must ask from the political as well as the "development" point of view, if there are any viable alternatives to such repair given that ethnic conflict saps national resources and have proven politically destabilizing. As D. Horowitz writes: "In divided societies, ethnic conflict is at the center of politics. Ethnic divisions pose challenges to the cohesion of states and sometimes to the peaceful relations among states...In divided societies, ethnic affiliations are powerful, permeative, passionate and pervasive" (1985: 12).

Kirby (1985) promotes a concept of the "local state" constituting a "coherent milieu for action-in-place at the subnational scale. It is an explicit political-geographic unit in which populations have repeated political struggles over many forms of issue." The local state is the point of encounter between civil authority and the political and social needs of ethnic groups. But there are other needs that must be met for constructive change to occur—such as the optimization of the relationship between a people and their environment, which are mirrored by the economic and ecological context of society. Endogenous Recovery Regions should capture the anthropological and ecological conditions that constitutes social habitat.

Critical theory (e.g., political economy, political ecology, Marxian analyses, the mode of production school, gender studies) provides necessary but insufficient tools for a complete understanding of ethnic unrest and entrenched poverty. Socio-spatial and geo-ethnic factors must be integrated into theory, planning and intervention if development is to succeed.

The Endogenous Recovery Regional Approach builds on Whittlesey's spatial conceptualization of the compage, Forde and Steward's culture area notion, aspects of the ecosystem and political ecology school, the complimentarity of indigenous knowledge and modern academic knowledge, the mode of production school's analysis relating to the articulation of local-level economies with those occurring at higher spatial levels, studies of gender-patterned social and economic life in a region, and the territorial approaches to regional designation assembled by Friedmann, Lo, Salih, Douglass, Weaver and others, and on Friedmann's and Forest's notion of the "politics of place" as the foundation for proactive regional planning. In this sense, the ERRGN approach can be understood as the evolutionary product of failed development as well as a re-conceptualization of society and its place in nature.

NOTES

1. Ethnic communities in exile or diaspora such as Armenians, Irish expatriates, Jews, and Palestinians may identify strongly with a territory they have never seen before. So, too, groups like Native Americans may inhabit areas (e.g., reservations) that they regard as constituting only a small part of their ancestral lands.

2. For example, the fragility of such consensus is sadly exhibited by the ongoing violence that has destabilized the social order between Singhalese and Tamils in post-colonial Sri Lanka.

3. The sectarian violence that transpired prior to and after Zimbabwe's independence between the predominantly Shona-speakers of the North (who were the political base of power for Robert Mugabe's ZANU-Patriotic Front party) and the Ndebele-speaking inhabitants of the South (who backed Joshua Nkomo's Patriotic Front-ZAPU party) suggests that caution be exercised in concluding that ethnic consensus has been achieved in Zimbabwe.

4. The term *ethnic segmentation* is used interchangeably with *ethnic discontinuities* and *fractionalization* to describe *ethnically divided societies* (i.e. *ethnic groups divided by borders* (e.g., Baluch), *multinational or multi-ethnic states* (e.g., Nigeria, Yugoslavia), *submerged nationalities* (e.g., Berbers), *persecuted minorities* (e.g., Moslem Turks in Bulgaria) and *non-state nations* (e.g., Kurds). The antonyms of these terms are *ethnic continuities, uniformities* or *contiguity*, which describe societies that are relatively unitary, from the ethnolinguistic point of view.

5. G.P. Nielsson (1985) differentiates between ethnic-groups and nation-groups on the basis of latent group identity that is present among ethnic groups whose autonomy is not challenged and nation-groups, which are ethnicities that have prosecuted political aims, particularly statehood, in response to challenges to their existence.

CHAPTER 6

CONTEMPORARY KURDISTAN AS AN ENDOGENOUS RECOVERY REGION

The collective existence of the Kurds has been problematic for all of the post-colonial states in which the Kurds live. The attempt of these states to impose monolithic social identities and the appearance of social unity on an ethnically diverse citizenry is an essential part of any explanation addressing the Kurdish "problem" (Bulloch and Morris 1992: 235). Assertion of Kurdish identity is viewed as undermining the Republican Turkish, Iraqi, Syrian and Iranian states, whose leaders seek to project the image of a unitary Turkish, Iraqi, Syrian, and Iranian identity upon their heterogenous societies. Beyond reasonable doubt, however, Kurdistan is a distinct ecoregion.

DIMENSIONS OF KURDISH IDENTITY

In characterizing the attributes of contemporary Kurdish identity across the colonial state borders that today divide Kurdistan,

N. Entessar writes: "[S]uch objective factors as race, language and religion have been less important in perpetuating Kurdish ethnonationalism than the subjective element of 'Kurdishness' based on a common way of life and a common historical experience. And for the Kurds, mountain dwelling has been the bond that has kept them distinct from the rest of society" (1989: 87). Entessar's definition of the Kurds emphasizes group solidarity based on the collective historical experience of a shared place—the mountains of Kurdistan—and continuities in social and economic patterns therein. Despite local cultural and linguistic variations[1] among Kurdish populations throughout Kurdistan, the rugged mountains of the region and the life it imposes on the people for whom it is a habitat constitute pillars of the Kurd's individual and collective sense of self (see Figure 6.1).

M. van Bruinessen (1983) discusses the political dimension of Kurdish identity. In the past the relative isolation of the Kurds gave rise to a political autonomy that is a vital part of Kurdish consciousness, albeit one strained at times by tribal rivalries. Kurdistan was a large frontier well-endowed with coveted resources, and Kurdish unity was more often displayed than not in the face of external Arab, Persian, and Turkish threat (Edmonds 1957; Malek 1989; van Bruinessen 1983).

The question, therefore, is not whether there is a discrete ethnonational group but, rather, what have been the consequences for the Kurds and the Middle East at large of their eclipse (Izady 1992; Kashi 1994; McDowall 1992).

DEMOGRAPHY OF THE KURDS AND OF KURDISTAN

Territorially, Kurdistan historically encompasses swaths of present-day Iran, Iraq, Syria and Turkey (Figure 6.2). Further, sizeable Kurdish communities can be found in Azerbaijan, Armenia, and Georgia, as well as in the major cities of Lebanon, Iran, Iraq, Syria and Turkey where Kurds live as rural migrants engaged mainly in menial labor. Communities of Kurds exist among the expatriate laborers working in European countries, particularly Germany.

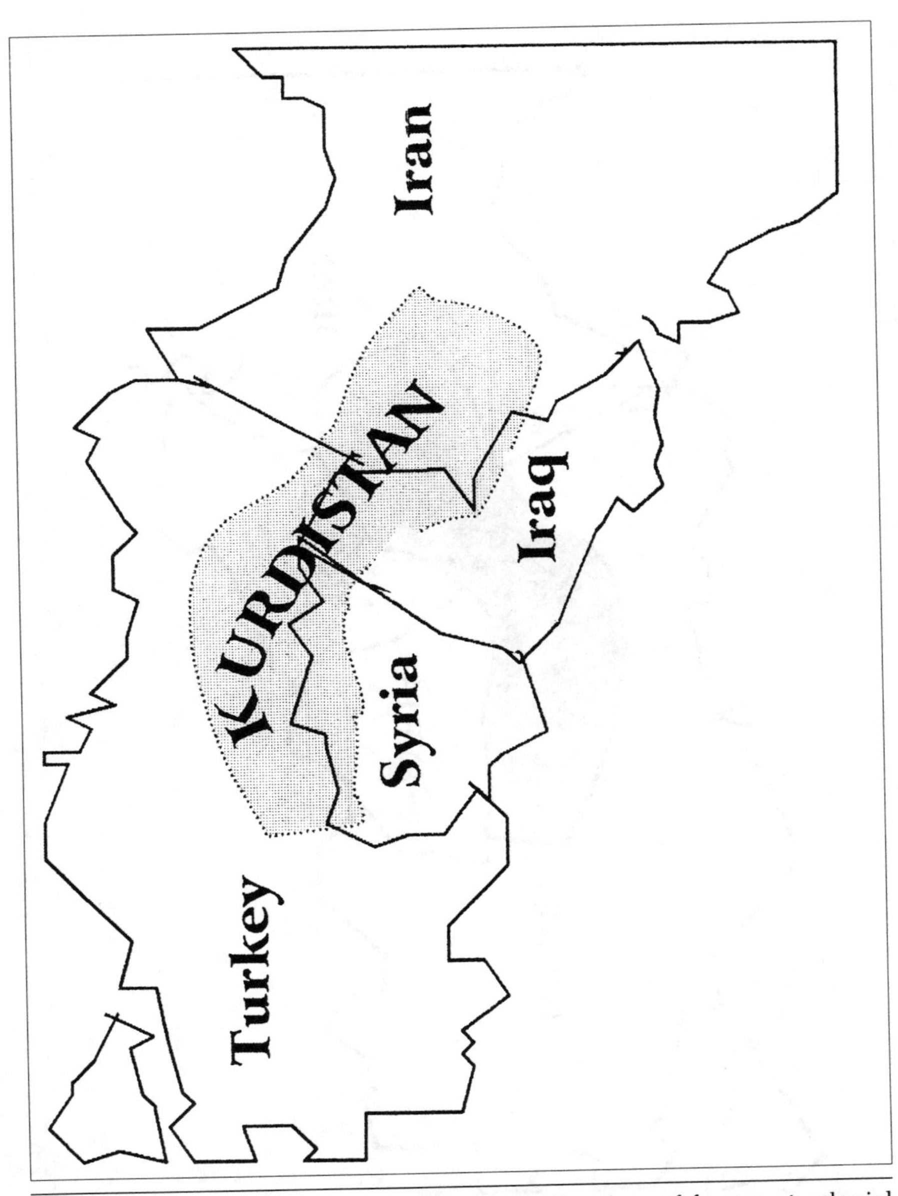

Figure 6.1 The division of Kurdistan by the borders of four post-colonial states. Note: The international boundaries of Kurdistan are unofficial. Accordingly, and with respect to projection and scale, this map should be taken as indicative, not representative.

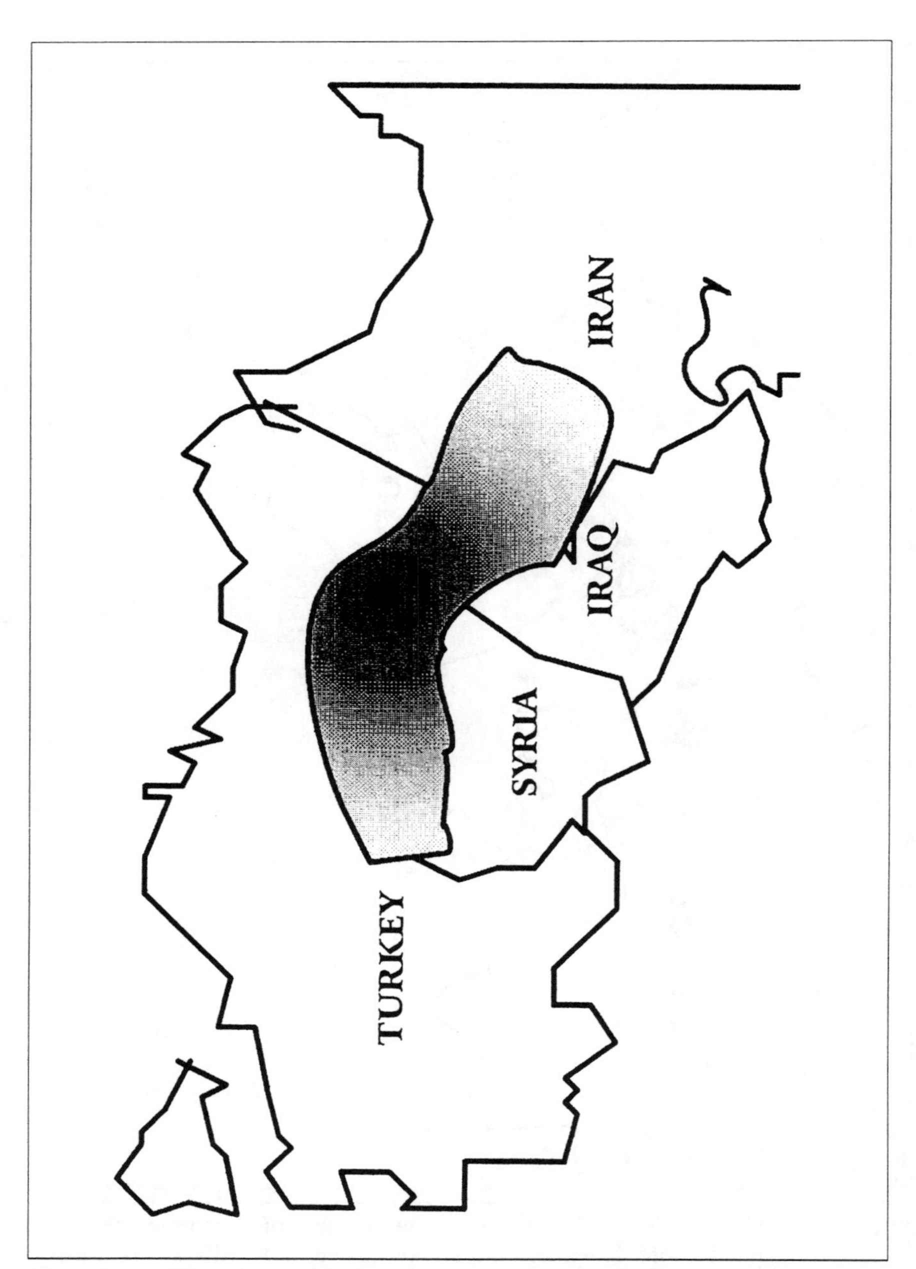

Figure 6.2 The relative numerical distribution of Kurds in Kurdistan. Source: Based loosely on McDowall 1989.

Ascertaining the number of people represented when speaking of "Kurds" is a difficult endeavor: "Nothing, apart from the actual 'borders' of Kurdistan, generates as much heat in the Kurdish question as the estimate of the Kurdish population. Kurdish nationalists are tempted to exaggerate it, and governments of the region to minimize it" (McDowall 1989: 7). Estimates vary from 14 million Kurds to more than 26 million (see Table 6.1) (Eickelman 1981; Malek 1989; McDowall 1989; Sim 1980). Most analysts believe that the higher values of this range more accurately reflect the true number.

Table 6.1 Estimates of Kurdish Populations, by Post-Colonial State

	Source and Year of Estimate				
	McDowall[2]	Eickelman[3]		Malek[4]	Sim[5]
	1980	1979		1989	1980
		Min.	Max.		
STATE	(in millions)				
Turkey	8.455	7.00	12.00	8.00	8.70
Iraq	3.105	2.00	3.00	4.00	3.00
Iran	3.701	4.00	6.00	5.00	4.50
Syria	0.734	0.50	0.60	0.60	0.60
(Sub-) Total					
Lebanon	0.060	0.12	0.07	0.10	0.10
ex-USSR	0.265	0.60	1.00	0.12	0.20
(Armenia, Azerbiajan, Georgia)					
Elsewhere	—	0.40	0.40	0.20	
Total	16.32	12.50	23.00	20.00	17.10

Taken as a whole, the Kurds are a major if obscured ethnic group in the Middle East: "Even though there is no agreement regarding the size of their population, it appears that the Kurds are the fourth largest ethnic group in the Middle East. Only the Arabs, Turks, and Persians, in that order, outnumber them" (Naamani 1968; see also MacDonald 1988; Sim 1980). Further, W.B. Fisher asserts that, "Even at minimum estimates, there are enough Kurds to form a 'state' that would certainly be larger than others now existing in the Middle East..." (1978: 195). D. Kinnane echoes this conclusion (1964: 2–3), as does M.H. Malek (1989).[6]

Kurdish society is an agrarian one (Ghassemlou 1980; Kendal 1980; Nazdar 1980). The rurality of the Kurds remains one of their most identifying characteristics. The states in which Kurds hold nominal citizenship have systematically exploited and denuded Kurdistan of its vast natural resources (Figure 6.3). Despite the natural wealth of Kurdistan, rural Kurds now encounter a depleted and scarred resource base.

Modernization, The Turkish State and the Kurds

Until recently the mere acknowledgment of Kurdish identity was sufficient grounds for a prison sentence in Turkey. Since the creation of the Turkish Republic, the Kurds have been brutally suppressed. Recent military campaigns in Kurdistan on both sides of the Iraqi/Turkish border have been justified in order to undercut the activities of the Kurdish Workers' Party (PKK) (Figure 6.4).

However, understanding Ankara's draconian approach toward the Kurds entails an appreciation of the economic relations between eastern Anatolia and the economic center of the Turkish Republic (Figure 6.5). Kendal's description of these relations is that "The flow of trade between Kurdistan and Turkey is, on the whole, quite in keeping with metropole-colony relations in general, where the colony serves the metropole as a reservoir of raw materials and as a protected market for its products" (Kendal 1980b).

Disparities between the Kurdish and Turkish sectors of the society becomes apparent when conditions in eastern (Kurdish) parts of the country are compared to those of the aggregate Turkish economy (Pittman, III 1988: Preface; 167). According to Seddon (1988), virtually all of Turkey's cash crops (e.g., cotton, tobacco) are grown in the Mediterranean, Black Sea, or Marmara-Thrace regions of the country. That is, the agricultural inputs for industry are concentrated in the West, not in Kurdistan where the crops grown are mainly for local consumption.

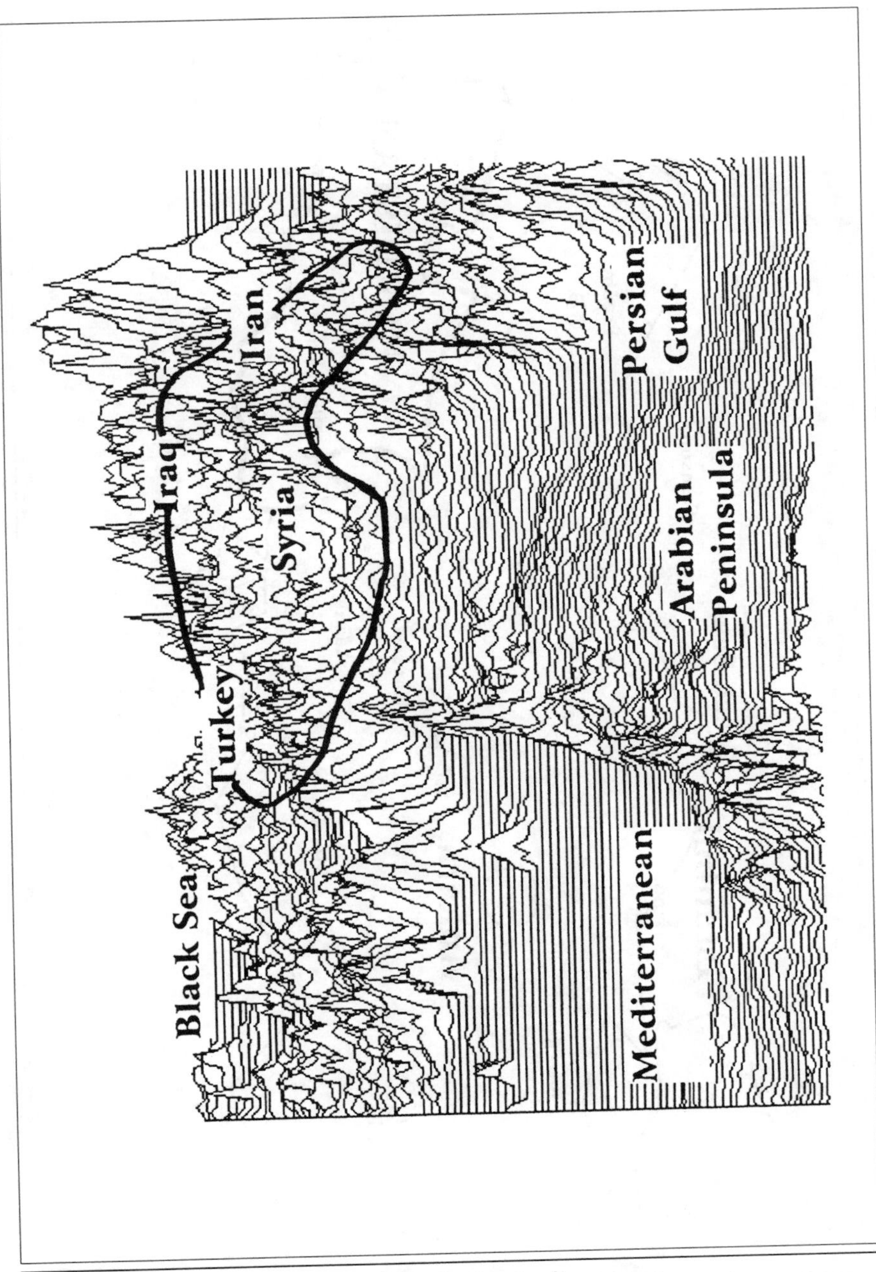

Figure 6.3 Terrain of Kurdistan and surrounding areas.

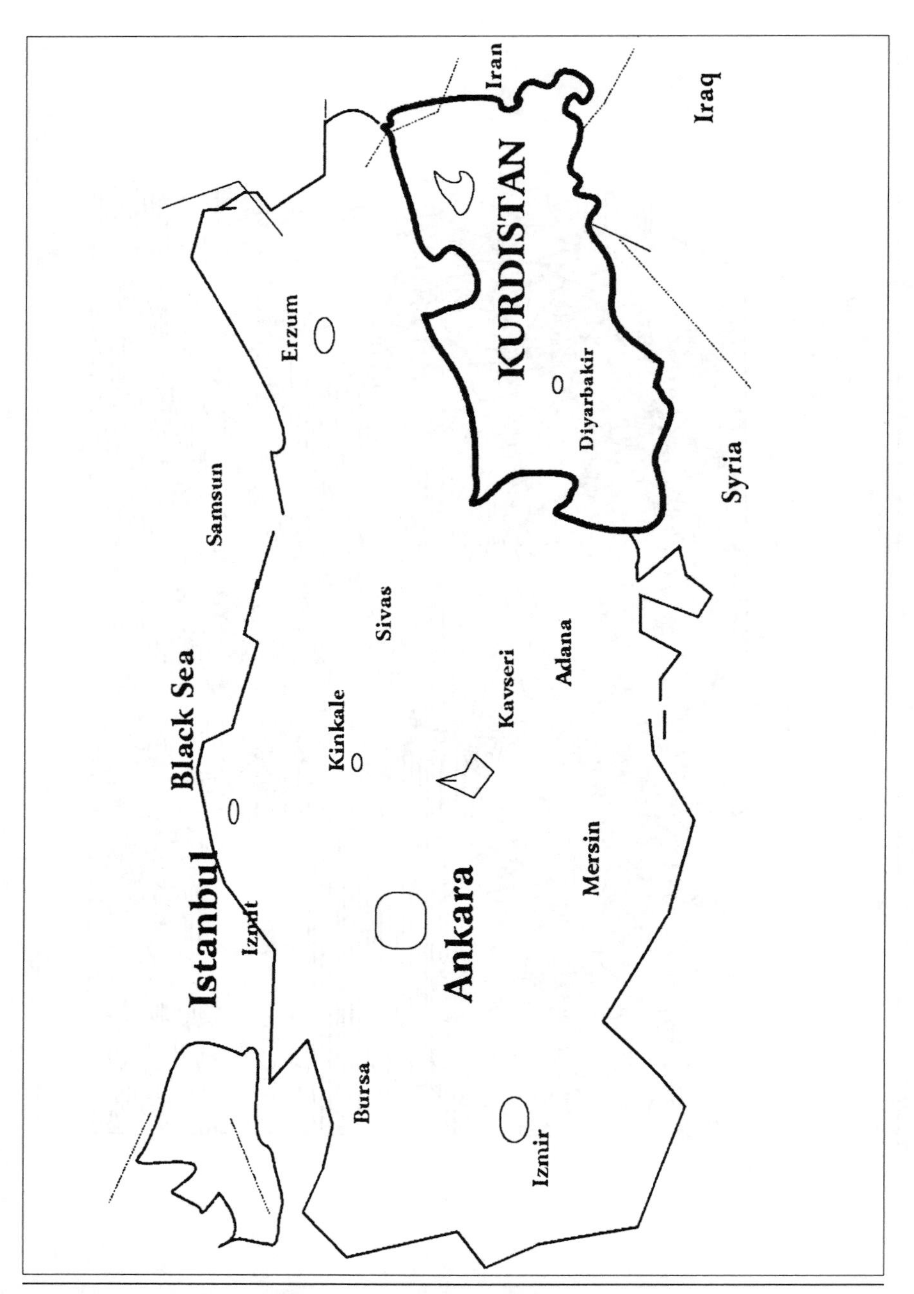

Figure 6.4 Kurdish regions in the Republic of Turkey.

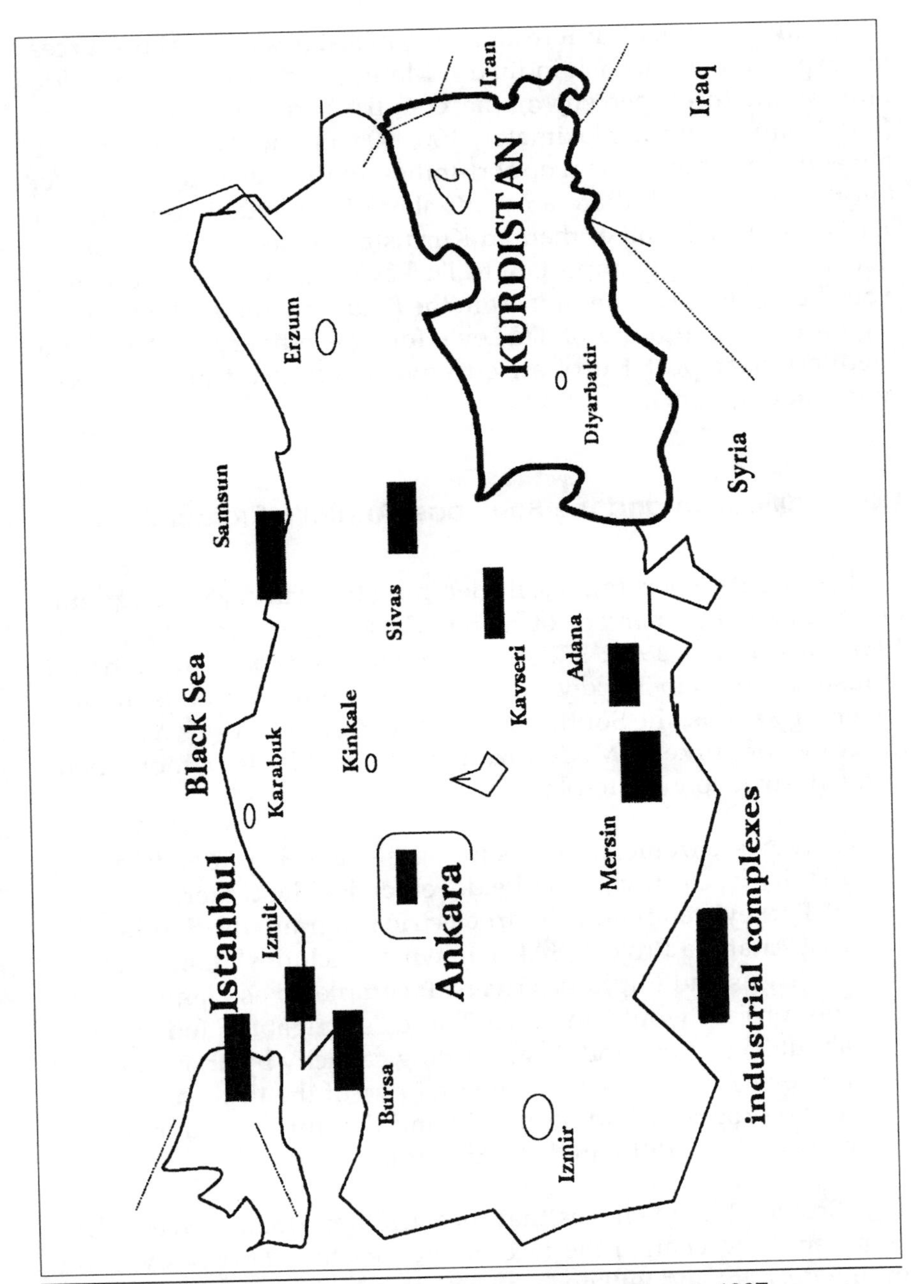

Figure 6.5 Turkey's industrial complexes. Source: Pitman 1987.

Unlike its agricultural resources, Kurdistan's mineral resources are exploited by Turkish industry. Mining and quarrying provide only about three percent of the country's total export earnings (ibid.), employing approximately 100,000 workers. These industries generate few employment opportunities for the Kurdish labor force. On the other hand, there are a number of exploitable oil fields in Turkey—virtually all of them in Kurdistan, as seen in Figure 6.6—that are increasingly important to Turkey's economy. Additionally, a 500 kilometer pipeline runs from the field centered in the Kurdish region of Batman (one of Turkey's four oil refinery centers) to a Mediterranean port. Kurds are effectively excluded from these economic sub-systems.

De-Kurdification and the Southeast Anatolia Project

Despite its emerging fossil fuel industry, Turkey's energy imports equalled 31 percent of merchandise exports in 1987 (Sunar 1987; World Bank 1989: 172). Given the role of Modernization and industrialization in Turkey's development strategy and its mounting energy needs, the Southeast Anatolia Project (SAP) is a jewel in the crown of Turkey's Modernization efforts. It is also a monument to the destruction of life-place.

> Turkey's chronic energy shortages made it imperative that hydroelectric power be developed....The centerpiece of Turkey's ambitious hydroelectric program, the Southeast Anatolia Project (SAP]), [...which includes] dams on the Tigris and Euphrates rivers. If completed as planned, this project would increase Turkey's irrigable land by about 25 percent and its generating capacity by about 45 percent. As of early 1987, the first two of the three dams in the program...had been built and the third was under construction (Pittman, III 1988: 214).

J. Brown (1988) believes that the 11 billion dollar project "has the potential to change the face of southeastern Turkey over the coming decades and will go far in assisting in the integration of the

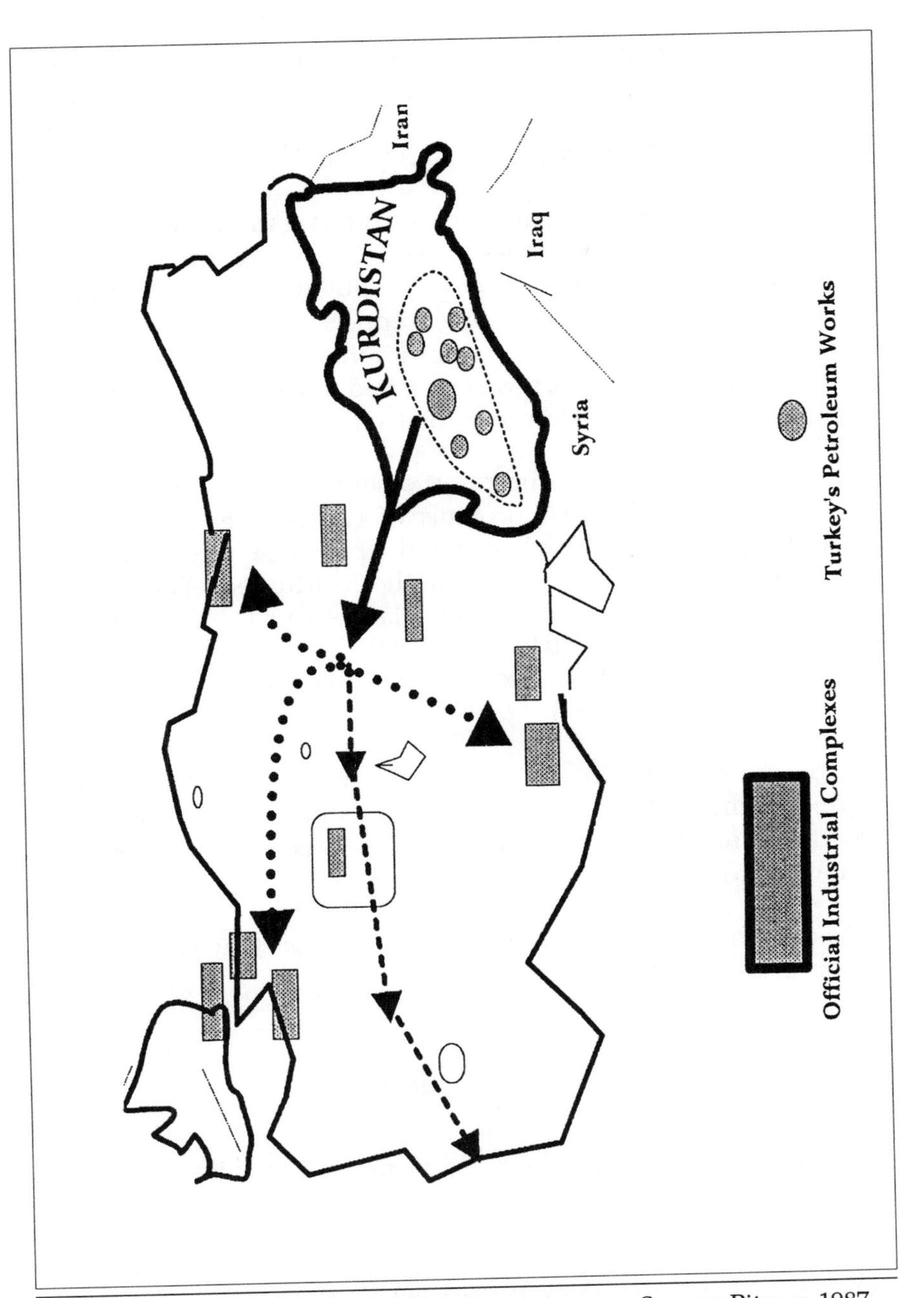

Figure 6.6 Location of Turkish petroleum reserves. Source: Pitman 1987.

Kurds, both economically and socially, with the rest of Turkish society." While D. McDowall speculates that the project "must have brought some benefit to the [Kurdish] area," the latter notes that "most of the hydroelectric power goes to western Anatolia" (1989).

As one of the major hydroelectric and irrigation projects in the world, the Southeast Anatolian Project entails significant restructuring of Turkish land use, ecology, population distribution, civil structures and economy. Prevailing skepticism regarding the project is justified given Ankara's record for implementing change in eastern Anatolia. For example, it was reported in 1987 that with respect to "development" in eastern Anatolia,

> the preparations for the resettlement of 234 of the 442 villages of the Kurdish province of Tunceli (Dersim) and of the 275 of the 559 villages of the province of Erzincan have been discussed in detail in the "controlled" Turkish press. In Tunceli 50,000 mainly Alawite Kurds have been affected and about 80,000 to 100,000 Kurds in the province of Erzincan have been resettled (the Gesellschaft fur Bedrohte Volker 1987).

One is hard pressed to view such massive displacement as "development" at least as far as the indigenous population is concerned. Obviously, this "development" has been done to meet the economic growth needs of the state center.

The situation in Turkish Kurdistan is consistent with that of Iranian, Iraqi, and Syrian Kurdistan.

Kurdistan in Iraq: Internal Colonialism

Vanly (1980) identifies the Kurdish provinces of Iraq as shown in Figure 6.7.

Accordingly, Kurdistan represents approximately 17 percent of Iraq's 438,446 km^2 (including the sparsely inhabited desert in Iraq's

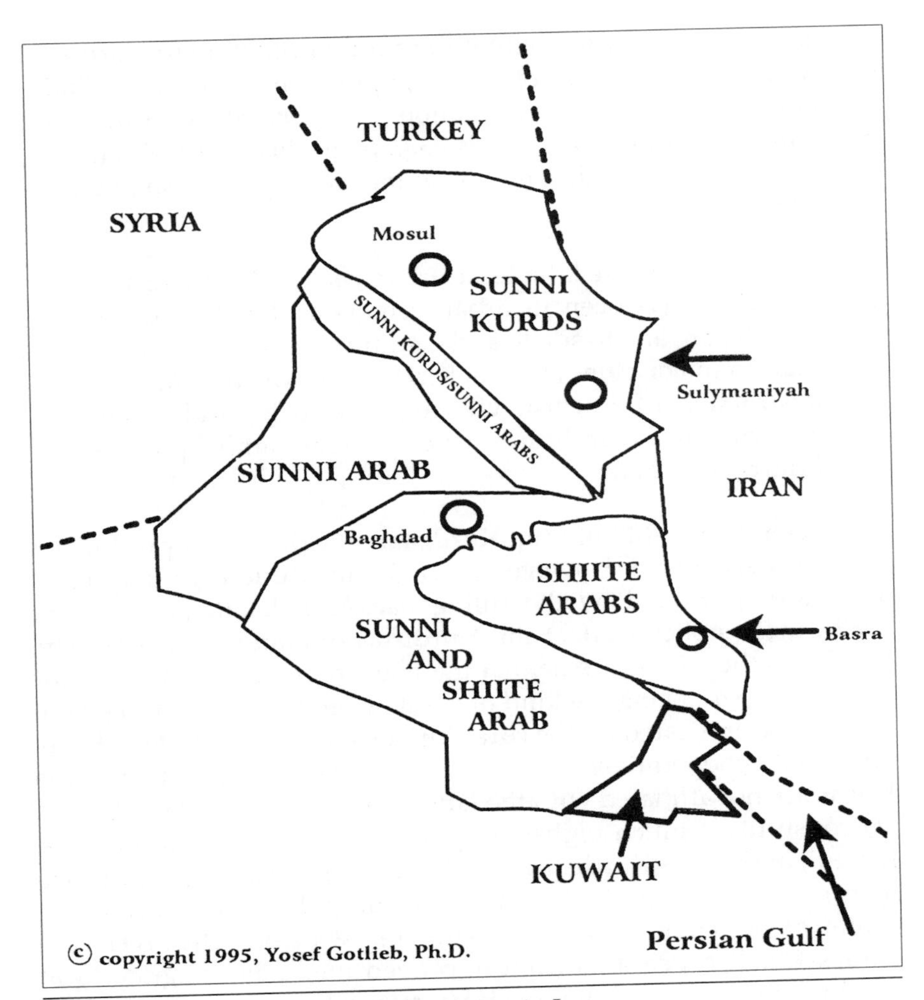

Figure 6.7 Ethnic groups and Kurdistan in Iraq.

southwest region). Using data compiled by the Federal Research Division of the Library of Congress (Metz 1990) from Iraqi documents largely unavailable to the author, we can determine the average density of Iraqi Kurdistan (consisting of the seven provinces defined by Vanly) to be 47.1 persons per km^2, as opposed to 37.2 persons per km^2 for Iraq as a whole.

The same sources show that the rural population in Kurdistan in 1987 was 2,081,000 out of a total population of 5,439,000 (ibid.), or 46 percent of the total population. The percentage of the rural populations in Iraq as a whole is 26 percent (Fisher 1989). Further, "urbanized" has very different meanings for the Kurdish and non-Kurdish sectors.

> Kurdish areas are...the least urbanized....[M]ost Kurdish migration has been to urban centers within the region [Kurdistan] and to small- and medium-sized towns, rather than to major cities such as Kirkuk and Mosul. War, the destruction of Kurdish villages, insecurity, and recent government expenditures have accelerated th[is] process (Marr 1985: 285).

Given the rurality of Iraqi Kurdistan, the dominant production sector is agriculture. However, since the institutions in Iraqi rural areas are controlled by the ruling pan-Arab Baath party (Stork 1982), these policies affect the Kurdish population deleteriously. Kurds, on the whole, are neither members of the Baathist party nor are they experiencing the kind of rural out-migration that the Arab sector is, as indicated by the rural/urban statistics provided above. Since state cooperatives are connected to the Baath, and since the Kurds are not affiliated with the latter, one assumes that the Kurds: (1) constitute a much higher proportion of the agricultural work force than the non-Kurdish sector (given their relative proportions in rural areas), and (2) the nature of the political and economic center of the country (i.e., the pan-Arabist Baath) effectively excludes Kurdish peasants from improved methods of agricultural production and, as I.S. Vanly (1980) states, for industry as well.

Oil and the Iraqi State Center

The importance of petroleum in the Iraqi economy and in fortifying the power of the state elite cannot be overstated (Stork 1982). In 1984 when the Iraq-Iran War was raging, oil, gas and related products comprised 85 percent of the country's exports (Metz 1990:

262) and generated an estimated 10 billion U.S. dollars annually. This amount rose to nearly 12 billion in 1985, dropped precipitously in 1986 (at the height of the War) but recovered to $11.3 billion in 1987 (ibid.: 261). It is important to note that the largest oil fields in Iraq are located in Kirkuk (Woodson 1975), that is, in Iraqi Kurdistan.

Iranian Kurdistan in Numbers

The three primary Kurdish regions in Iran consist of the *ostans* (provinces) of West Azerbaijan, Bakhtaran (Kermanshah) and Kordestan (Figure 6.8).

The Plan and Budget Organization of Iran (1984) represents their collective population as shown in Table 6.2.

In describing the inequities relating to the Kurds and other minorities in Iran during the later years of Pahlevi rule, E. Abrahamian writes:

> ...the regime's economic and social programs tended to increase regional inequalities. For example, Tehran obtained many of the new assembly plants and over 60 percent of the loans given by the Industrial and Mining Development Bank. Consequently, by 1975 Tehran produced half of the country's manufactured goods and contained 22 percent of the country's industrial labor force. In Tehran, for every worker employed in manufacturing there were 0.7 in agriculture. But in East Azerbaijan the ratio was 1:2.6; in West Azerbaijan [Kurdistan] 1:13; and in Kordistan [Province] 1:20. Similarly, the literacy rate was 62 percent in Tehran, but only 27 percent in East Azerbaijan, 26 percent in Baluchistan and Sistan, and 36 in Kordistan. Tehran had one doctor per 974 people, one dentist per 5,626 people and one nurse per 1,820 people. On the other hand, East Azerbaijan had one doctor per 5,589 people, one dentist per 57,294 people, and one nurse

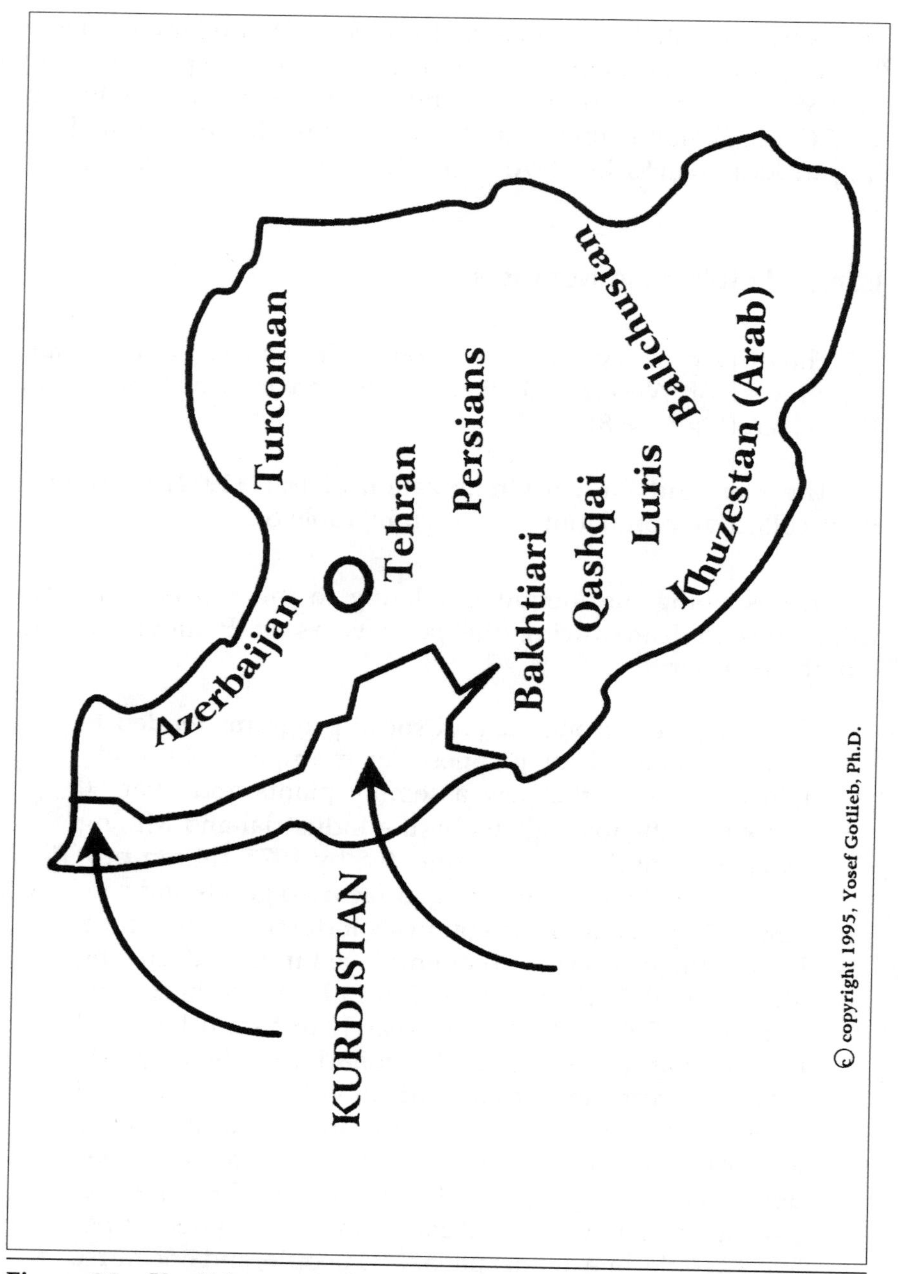

Figure 6.8 Kurdish areas in Iran.

Table 6.2 Population of Kurdish Provinces in Iran, 1983–1984[a]

Province	Population (in 000s)	Area km²	Density Pop./km²
W. Azerbaijan	1688.0	38850.0	43.5
Bakhtaran	1225.0	23667.0	51.8
Kordestan	923.0	24998.0	36.9
Iranian Kurdistan	3836.0	87515.0	43.8
Iran—Total	37700.0	1634958.0	23.0
Kurdistan as % of Iranian Total	10.2	5.4	190.0

[a] Corresponding to the Iranian Year 1362
Source: Plan and Budget Organization, Statistical Center of Iran 1984

> per 46,552. Finally, Baluchistan and Sistan had one doctor per 5,311 people, one dentist per 51,663 people, and one nurse per 27,064 people. The resentments built up by these ethnic and class inequalities remained hidden during the early 1970s. But once cracks appeared in the Pahlevi regime, they rushed forth in a torrent to engulf the whole society (1982: 449)

Data assembled by Amirahmadi (1987) indicates that there is a considerable gap in the number of health care providers between the Kurdish and non-Kurdish areas in Iran; further, the relative difference between the national average and the Kurdish regions for doctors actually increased after the Islamic Revolution. With respect to the number of hospitals and hospital beds, there are nearly one-third and two-thirds fewer hospitals per capita and hospital beds per capita respectively in Kurdistan relative to the state average.

Kurdish areas also lag behind non-Kurdish areas in literacy rates and percent of change of literacy rates. Further, Ministry of Education and Training data as reported in the Plan and Budget Organization's Statistical Center of Iran 1984 show that Kurdish areas lag behind non-Kurdish areas in school enrollment at all levels of education.

More recent occupational statistics by province have not been available, although H. Amirahmadi (1987) assembled this data for 1956, 1966 and 1976. For all three periods the Kurdish regions consistently demonstrated employment structures in which the agricultural sector dominated the other sectors. Also, the share of agriculture in the labor force was almost always larger in Kurdish regions than in the non-Kurdish areas for the corresponding periods. This is paralleled by inequities in the distribution of development funds by province (Amirahmadi 1987, employing Plan and Budget Organization data contained in the *Fifth Development Plan*, 1968–1977, published in 1972, p. 127).

Amirahmadi has also compiled data showing that during the period from 1961 to 1978, Iranian Kurdish areas received 2.9 percent of the total loans and investments provided by the Industrial Mining and Development Bank of Iran (IMDBI) for a total of 12 loans or direct investments out of 774 issued in Iran during that period; one Kurdish province, Kordestan, did not receive a single loan or investment from the IMDI. West Azerbaijan contains 4 out of 34 mines and quarries in Iran yet received government assistance well below what one would expect given its productive capacity (Plan and Budget Organization 1984).

In social and economic indicators alike, available statistics show that Kurdistan was consistently behind the Iranian state average. This remains true even after the fall of the Pahlevi dynasty and the establishment of the Islamic Republic. Further, if statistics for the Parsi (Persian) provinces (encompassing about 60% of the population) were differentiated from the national average, it is clear that the data for the Kurdish and Persian sectors would show an even larger disparity.

KURDISTAN AS AN ECOREGION

The above survey of historical and contemporary conditions affecting the Kurds lead one to the following conclusions concerning Kurdish social ecology:

- *Historical:* Anthropological and archeological evidence document Kurdish history as extending at least as far back as the ancient Medes period. The Kurdish language is distinct from all others, is spoken in two primary dialects, and has been the vehicle for "social communication" (Deutsch 1966) among the Kurds. Their traditional social organization is the tribe. Seasonal semi-nomadic herding was traditionally practiced by many Kurdish tribes. The Kurds are predominantly Moslem, with Sunnism being the dominant denomination, although various Shiite sects, Christian groups, and other creeds (e.g., Yazidis) are found in Kurdish communities. Even when under the nominal jurisdiction of regional powers, the Kurds have maintained internal autonomy.

- *Ecological:* While there is local variance in the ecology of Greater Kurdistan regional continuity exists in terms of geomorphology (rugged terrain characterized by mountainous areas and valleys) and of climate (northern temperate to sub-alpine). Although landlocked, Kurdistan has abundant sources of fresh water. Geographically, Kurdistan is a crescent-shaped slice of territory constituting approximately 500,000 km^2 and located within the area defined by 37.5 to 47 degrees from West to East, and 39 to 34 degrees from North to South, excluding much of the southeastern quadrant of the area demarcated by these coordinates.

  Since the post-colonial state centers controlling Iran, Iraq, Syria and Turkey have adopted "scorched earth" military policies in attempting to put down Kurdish insurgencies, environmental degradation has accompanied the subjugation of the Kurds (Kashi 1994; Nelan 1992). Consequently, the application of indigenous knowledge to the territory has been arrested; resources have been commodified and diverted from the Kurdish periphery to the state core, and varying policies have been pursued in the unitary Kurdish ecoregions that now fall under the authority of the different states to which Kurdistan has been apportioned.

- *Ethnic Identity:* Kurdish group identity is greatly influenced by a common ethno-habitat, history and mode of production. Historical isolation in their mountain principalities left a strong tradition of independence among the Kurds. A political culture leading to an accommodation with the dominant power ruling them maintained Kurdish autonomy in return for nominal allegiance to regional hegemonies. When such autonomy was threatened, Kurdish history reveals that rebellion was the general response. A cross-border Kurdish nationalism emerged during the twentieth century led at first by tribal chieftains and later by Kurdish military men who had formerly served the Ottoman or Qajar Empires or their successor states. Increasingly, intellectuals have assumed the leadership of the Kurdish national movements, which have generally been politically leftist during the past half-century. Multiple political parties exist, most covertly, throughout Kurdistan. Many of these parties have Kurdish fighting forces (*peshmerga*) associated with them, although a united Kurdish Front has emerged of late. The parties represent conflicting ideologies and/or economic (classes), strategies, and power interests within the Kurdish population.

- *Economic System:* The traditional mode of production that continues in some rural Kurdish settlements remained based on semi-nomadic animal husbandry and a seasonally sedentary agriculture of staple crops, fruits and vegetables. Historically, relations of production have been feudal and distinguished by a numerically large class of commoners working lands belonging to the tribal *aga*. Most frequently, economic relations with ruling imperial states (e.g., the Safavids, Ottomans, Qajars) was conducted through the *aga* and the state agent. Tribute was paid by the former to the latter on behalf of the tribe.

Tribal structures have increasingly faded and have been replaced in importance by political affiliation. Oppression against the Kurds has been far less vertical due to class stratification and more horizontal owing to the oppressive policies of the state elites that rule over Kurdistan, a triumph one presumes of the influence of left-leaning parties in the Kurdish movement and an expression of the oppressive regimes that affect the different Kurdish classes simultaneously, albeit differently. For example, land ownership is no longer an internal Kurdish matter but is determined by ruling state policies. The number of Kurds with appreciable investable capital is minimal, although there is a peripheral Kurdish mercantile class. The latter generally appears to comply with the aspirations of the Kurdish movement as a whole.

- *Underdevelopment:* Historically, feudal structures rendered Kurdish society hierarchically stratified, but not intricately so. Most Kurds were commoners, and the privileged classes of agas, sheikhs and religious functionaries was numerically small. With European economic penetration of the Middle East, the Kurdish mode of production was gradually distorted by changes in agriculture, increasingly oriented from subsistence to cash-cropping. This distortion was induced by a (generally non-Kurdish) comprador class that served as the link between the traditional Kurdish economy and the increasingly extroverted economy of the Ottoman and Qajar states during the nineteenth and early twentieth centuries. The products sold were primary commodities such as tobacco and preserved fruits.

  With the introduction of new states into the post-colonial Middle East and the division of Kurdistan into four parts under non-Kurdish jurisdiction, core-periphery relations defined the interaction between Kurdish regions and the states of which they are a part. Material resources and labor are extracted from the Kurdish areas without any meaningful reinvestment for development purposes. Accordingly, the transfer of Kurdish resources has steadily broadened the already wide socioeconomic gap between Kurds and non-Kurds in the post-colonial states. Among these resources are petroleum and mineral ores. The post-colonial state economies have essentially captured the Kurdish regions. Kurds have been used as a reserve labor force for menial, labor-intensive tasks both within the states and as workers exported abroad for revenue-generating purposes. The Kurds serve as a reserve army of labor directed by the state (through the demands of the labor market) and driven by economic necessity: Traditional subsistence patterns are no longer sufficient to maintain a household. The result is consistent with patterns of internal colonialism and "irreconcilable planning" (Gotlieb 1996).

- *Patterns of Oppression:* As discussed throughout this chapter, the incorporation of Kurdistan into the post-colonial states of Iran, Iraq, Syria and Turkey has accelerated the underdevelopment of the Kurds that began during the waning decades of the Ottoman and Qajar empires. The Kurds have been denied autonomy, national minority rights, and, often, civil rights. All of the post-colonial states governing Kurdish territories are highly centralized, ethnically stratified polities with oppressive internal security apparatuses that ruthlessly enforce state elite power (McDowall 1992).

De-Kurdification (Malek 1989) in the form of either pan-Turanianist, pan-Arabist, or Islamic fundamentalist identity and forced assimilation has been shown to be a common policy among the state elites governing Kurdistan. Invariably, this has led to persistent Kurdish resistance. Strategies employed to subjugate the Kurds include prohibitions against the use of Kurdish language, dress, and culture; forced resettlement and deportations; detention and torture of political prisoners or suspects, often without trial; the imposition of martial law over and the occupation of Kurdish areas by state security agents empowered to act by fiat; land expropriations; forced conscription, including that of adolescents; military actions against civilians as well as *peshmerga*; defoliation of Kurdish lands; the absence of or severe deficiencies in the provision of state services in Kurdish areas; and attempts at genocide through the use of nerve gas and other weapons of mass destruction.

The primary motivation of policies applied to the Kurds by governing elites entails the complete territorial integration of Kurdistan in the post-colonial states and the consolidation of the region's economic and strategic assets as state (i.e., elite) property.

Given the definition of the Endogenous Recovery Region previously introduced, it is clear that Kurdistan constitutes a distinct Endogenous Recovery Region—historically, ecologically, anthropologically and by virtue of a common experience of group oppression (Gunter 1988).

NOTES

1. The Kurdish language is spoken in two major Kurdish dialects; Zaza (or Kirmanji), which is spoken in northern Kurdistan (Anatolia), and Gurani (or Kurdi), which is used in the southern region (Mesopotamia and Persia). It is a unique language related to but distinct from Persian and is used by no other ethnic groups except the Kurds. For a more detailed description, see Arfa (1966).

2. McDowall, D. 1989. *The Kurds.* (Report No. 23). London: Minority Rights Group, p. 7.

3. Eickelman, D. F. 1981. *The Middle East: An Anthropological Approach.* Englewood Cliffs, NJ: Prentice-Hall, p. 165.

4. Malek, M.H. 1989. Kurdistan in the Middle East conflict. *New Left Review.* Vol. 17: May/June, pp. 79–96.

5. Sim, R. 1980. *Kurdistan: The Search For Recognition.* Conflict Studies. No. 124, November.

6. Although they are often referred to as a "minority" group within Turkey, Iraq, Syria, and Iran, the Kurds are an overwhelming majority in that large expanse of land in northwestern Asia that had, until recently, been known as Kurdistan (MacDonald 1988: 195).

CHAPTER 7

ENDOGENOUS RECOVERY AND A CHOICE OF FUTURE(S)

The Kurdish situation serves as a case study that illustrates the development of global dysfunction. There are differences within Kurdistan relating to variations in colloquial Kurdish language, tribal affiliation, ideology and degree of physical isolation. Nonetheless, Kurdish "within" group differences relative to the "between" group differences among Kurds and other groups pale in significance. Historically, the Kurds have had a distinct life-place, socio-linguistic composition, culture, adaptive mode of production and autonomy. As a nation of at least 25 million people, the Kurds struggle to maintain control over their lives despite the processes of Modernization that have obscured and disinherited them.

Kurdish life-place has been badly mangled by jet bombers and resource diversion. Radical changes have been applied to every dimension of Kurdish life-space: *economically*, as in the great gaps pertaining to resource expropriation and economic infrastructure of the Kurds relative to non-Kurds; *ecologically*, as in the cases of the Southeast Anatolia Project and the defoliants and napalm dropped by military jets in Iraqi and Iranian Kurdistan; *socially*, by the forced resettlement and "Arabization" of the Kurds displaced from Syria's Jazireh Strip, and; *culturally*, especially in Turkey where identifying oneself as a Kurd leads to discrimination and, until recently, summary imprisonment.

Despite all this, Kurdistan and the Kurds continue to exist largely due to the vitality of the people and their determination to live freely on their own land. It would appear that Kurdistan as an Endogenous Recovery Region holds far more promise for progressive socioeconomic change and environmental sustainability than its division under post-colonial borders would allow. Further, Kurdish autonomy is a moral imperative as well.

SOCIAL CRISES

The world knows little of the Kurds—or of the Tibetans, the East Timorese, the Berbers, the Karens and scores of other submerged peoples around the globe. Some of these groups are numerically minute and on the verge of extinction while others like the Kurds number in the millions. Despite their obscurity their plight is no less severe than that of the absolute poor of Somalia, Nepal, Rwanda or Haiti, where meeting basic needs is a day-to-day struggle. For both the absolute poor and members of submerged nationalities—to be sure, the two categories are not mutually exclusive—the dream of a better life is as torturous as malnutrition and as horrendous as the fresh grave containing a child's bullet-ridden body.

Contemporary social crises are not restricted to post-colonial or other Third World societies. In what was once termed the Second World—the countries of the former Soviet Union and the Eastern Block—rampant crime, corruption, alcoholism and political extremism have reached nearly pandemic proportions. The fall of Soviet totalitarianism seems of little solace to hundreds of millions of people whose lives are as dismal as ever. Throughout the former

Soviet Union, ethnic rivalries as old and as bitter as those that have recently torn through Chechniya have intensified. Former brothers-in-arms of the Red Army have redirected their gunsights from abroad to within. Aside from human losses in many areas of the former Soviet Union, the wages of war will be bitter given the ecological disasters that have already taken place and those that are in the making.

But perhaps the most jarring of global dysfunction's manifestations are found neither in the post-colonial or the former Soviet Bloc. The alienation and disaffection encountered in the "advanced" industrial countries are signs of Modernization's destructiveness. Human beings have become "objects," specifically consumers, and the accumulation of "things" is the penultimate value uniting western civilization. Representative of this ethos are television commercials which would make us believe that owning a certain vehicle or consuming a given packaged cereal are life and death issues. Squeezed between the glorified violence or silly pretensions of most commercial television programs, these advertisements are part of a superstructure that exhorts us to consume, consume, consume. Similarly, children living in urban areas are to draw their inspiration from the pageantry of modern gladiators engaged in hyper-capitalized "sports" and entertainment. Still the great majority of minority children, and increasingly the children of mainstream America as well, see violence, drugs, unemployment and suicide as their most accessible options. Anti-social behavior can be termed "adaptive" in a world where work, responsibility and scholarship do not constitute shields against drive-by shootings, carjackings and domestic violence.

The Loss of Wisdom, The Loss of Self, The Loss of Community

Increasingly, displays of fanatical rage and ennobled extremism have become commonplace. A suicide bomber walks onto a bus in Tel Aviv, detonates the explosives beneath his clothing and kills two dozen people along with himself. American "super-patriots" declare "war" on their own government by placing a bomb of incredible force under a day-care center in the federal building in Oklahoma City. A Christian fundamentalist minister defrocked by his Church openly admits in court to having shot a physician to death for performing legal abortions. A mother in North Carolina claims that her two sons were kidnapped only to confess a week later to having killed her children and concocting the story. Hutus and Tutsis play out their rage by slaughtering each other's minorities in Rwanda and Burundi. The Sublime Truth cult attempts to impose its doomsday prophesies on Japan by using nerve gas on thousands of subway commuters. Cultists in Waco, Texas, in Switzerland and in Québec commit suicide (some with a great deal of "assistance") rather than face a life free of their leaders' tyranny.

These and multiple other social crises reveal a humanity that is rapidly loosing its soul. Our jaded Modernism has created a culture of not feeling. We are offered and have come to prefer being "comfortably numb" in return for complacency.

A COSMOPOLITAN WORLD

The world had become accessible to the monied minority of its citizens. This economically dynamic elite insures that places on every continent—Bangkok, Nairobi, Cairo and Buenos Aires—have the same fast-food restaurants, hotel chains and cosmetic products available to them in Paris, Milan or New York. Increasingly the world is one place, tinsely and technic, metropolitan and global to the extent that one can cross thousands of miles by air, disembark, and encounter the same landmarks of social reality as those at "home"—Hiltons and Sheratons, Visa and American Express, chain eateries and designer clothing. Exotica is local color buffered by the uniformity of conditions to which Northern tourists are accustomed. Place has become disposable and interchangeable.

Global capital has packaged nature so that it is sufferable and even entertaining in the context of weekends in Aruba, ski trips to Aspen and the Alps, and "safaris" to Kenya. Airlines have scheduled hundreds of flights each day linking both "civilized" and "primitive" nodes of economic space. Communication by telephone, teleconferencing, electronic mail and facsimilies make the interchange of information a totally "place-less" endeavor. The implications of placeless-ness is that human and geographical landscapes are marketable objects detached from their social substratum and the physical setting that nourishes them.

Were it possible for all of humanity to enjoy the "good life," a case could be made for encouraging the growth of market economies. But the social limits to growth *do* appear in the form of the gap between rich and poor, the disparaties in the numbers of people who have access to resources and services and those who do not, and in the social alienation represented by religious fanaticism and nihilism.

There is, of course, a second hemisphere of the "single social reality"—the ecological—and the limits to growth that it imposes appear to be just as potent (and probably more so) than social limits to growth. Nature's cybernetic mechanisms seem much more resilient—and unforgiving—than our attempts to subjugate our environment would suppose.

ENVIRONMENTAL DEGRADATION

The October 1994 oil spill in Russia's Komi republic is estimated to have greatly exceeded, perhaps by an order of magnitude, the amounts released in the Exxon Valdez incident. Even if it proves considerably less catastrophic in scope, it took place in a critical zone—the territories and waterways adjacent to the Arctic icecap. Oil is reported to have filtered perilously near the Arctic Ocean. Had this happened, it is conceivable that climactic conditions over the polar cap would be affected, at least temporarily. The Arctic's central role in the global hydrological cycle would undoubtedly be disturbed, and increased sea levels along the Canadian Maritimes, the northeastern coast of the United States, and along the coasts of Scandinavia and northern Europe, as well as Russia, would be likely.

The rise in sea level would be indiscernible, and it would likely take a season or more before the thermal effects were detected. Still, a likely impact area would be the Gulf Stream, the current of warm water that swerves off the coast of Florida into the northern mid-Atlantic. A departure of even one degree could induce thermal shock among fragile plant and animal species in the marine ecosystem. Were this to occur, species higher on the food chain would be affected. The extent of damage could not be predicted in advance.

The crisis described above may not happen. One could justifiably argue that the hypothetical conditions would not materialize and, if so, they would have only transient effects. However, in the event of a negative outcome from such an incident, the impact could be much more complicated and devastating than we could anticipate on the basis of current knowledge.

We simply don't know.

And that is a major characteristic of global dysfunction—uncertainty. We have been so impressed by what we have discovered that little attention has been paid to how deficient our *understanding* is of potent technologies..

Whether out of myopia, intoxication with our power to unleash tremendous force and change, or greed for what riches might result from the progress of Modernization, we have been truly blind to the ecological and social impacts of hypergrowth and life-place displacement. Only now are we beginning to comprehend our ignorance. There is an emerging awareness of how much we do not know and how high the stakes are in the adventure of Modernization.

What We Do Know

We do know that fishing boats working out of New Bedford, Massachusetts and Portsmouth, New Hampshire must travel further out into Georges Bank to find whatever catch can still be culled from the once thriving sea. The thinning stocks in home waters have compelled captains to venture further in search of more abundant fishing grounds. American and European vessels been boarded by the Canadian Coast Guard after entering Canada's territorial waters and their catches impounded. There are Canadian fishing families who also must be fed. The barrenness of the sea, like acid rain, knows nothing of national boundaries.

Recently, the Central American isthmus has been gripped by drought. It is thought that the unusual lack of precipitation is a consequence of climatic disturbances in a major current off of South America. El Niño's cyclical tempering of large expanses of water is a major force in the ecosphere. Climatologists throughout the world are formulating models of why El Niño has become so intemperate. Some attribute it to the fires set to clear vast tracts of ground cover throughout South America, while other modelers feel that the rapid deforestation of the Amazon basin, the hemisphere's most crucial carbon sink, is the more likely reason for El Niño's recent erratic performance.

Elsewhere malaria, tuberculosis, yellow fever and even the bubonic plague are reemerging and proving resilient to the antibiotics developed by microbiologists for use against earlier strains of the pathogens. In Africa and elsewhere communicable diseases, rural-urban flight, and war are decimating villages whose population pyramids are already swollen at their base due to high dependency ratios. Under such conditions hunger is chronic, and mortality rises among women and children who spend larger portions of their days in search of fuelwood and drinking water.

The fuelwood crisis which derives from population pressure has resulted in large amounts of land becoming stripped of cover. Soil erosion, sedimentation, blocked rivers, water scarcity and further disruptions in the symbiosis of flora and fauna is the unavoidable result. Experience shows that soils will be more intensively cultivated and that increasing amounts of rangeland will be grazed bare by emaciated cattle herded by emaciated people whose growing desperation is moderated only by their waning strength. The process, we have learned, is cyclical and cumulative as G. Myrdal (1957) showed to be the case in the socioeconomic sphere.

HISTORY RECONSTRUCTED

The Komi oil spill, changes in the El Niño current, the fuelwood crisis and numerous other environmental stresses are among the early signs of global ecological dysfunction. Manifestations of ecological collapse are predicted to take place with increasing frequency and duration. They will be accompanied by the social and economic phenomena of dysfunction described above.

The Modernization trajectory of social change can be understood[1] as a matrix wherein different periods of history correspond to distinct *Modes of Production, Horizon* (spatial and temporal limits of consciousness), *Habitat* (perceived situation of life-place in nature), *Labor* (justification for work) and *Ecological Consciousness* (consciousness of the relationship between society and nature), as depicted in Table 7.1

Correlated in this way the interrelationship between societies and their environments are seen as profoundly different prior to and after the agricultural revolutions. In pre-agrarian societies humans perceived themselves to be an integral part of a larger whole, the life-place that surrounded them, and which also circumscribed the spatial limits of consciousness, their sense of "home," and their economic relations (primarily undertaken for subsistence needs with little or no intervention in natural processes). The predominant world-view was one of fusion with the natural world.

With the Agricultural Revolution a fundamentally different consciousness emerged. Nature and society were still viewed as closely interwoven but also distinct from each other. Land was something "of" nature that was claimed by a privileged elite, e.g., a monarch, the feudal aristocracy, the church. This suggested an important dualism with society and nature divided into two interacting realms. Human labor was invested in agriculture not only to meet subsistence needs but also to generate a surplus paid in tribute to the elite. The latter's domain of governance "contained" the village to which people "belonged." The social consciousness was rooted in authoritarianism justified by a legitimizing creed (religion,

Table 7.1 The Evolution of Ecological Consciousness

Historical State	**Mode of Production**	**Ecological Consciousness (Society-nature relations)**	**Horizon**	**Habitat**	**Labor**
Pre-agrarian	Subsistence; Steady-state	Virtual identification World-view: *Intuitive; Fusion*	Ecosystem	Ecosystem	Ecological
Post-agrarian	Agriculture; Slow growth	Cultivation of "Land" for human needs; Emergent dualism World-view: *Inspired Authoritarianism*	Village to regal domain	External to ecosystem	Basic needs plus tribute
Industrial	Mechanization; Accelerated growth	Full dualism Removal of natural impediments on expanding scale; Subordination of Nature World-View: *Scientific Positivism* *Post-place*	Town/country to inter-statal	Human place out of nature	Amenities driven (Differential by class)
Post-Industrial Dysfuntional	Automation Production of Needs; Consumptionist Hypergrowth	Irrelevant or alien/social narcosis Poverty, hypergrowth, degradation World-view: *Relativistic;* Uncertainty, fragmentation *Post-Social/Post-nature* *Cyberspace*	Urban to Global to Artificial Nature (i.e., Virtual Reality)	Inter-Changeability and Disposability of place	Fractal, Aimlessly Baroque

aristocratic descent, mythology). This period involved slow growth with agricultural surpluses used to trade with or fuel the conquest of other peoples.

Formally speaking, one can say that the Modernization trajectory had its roots in the Agricultural Revolution since the latter necessarily intervened in environmental processes, albeit in a small-scaled and generally low-impact way. However, agricultural production even of a most rudimentary sort entailed an attempt to control aspects of nature (irrigation, planned cultivation) that was qualitatively different than the philosophical, deliberate endeavor of the Industrial Age to subjugate nature completely. The Enlightenment had given birth to scientific positivism, a way-of-knowing that considered itself progressive and irrefutable. It would eradicate ignorance and all things primitive. It was during the Industrial Revolution that Modernization began in earnest.

Technology and Liberation from Nature

It was no accident that European imperialism coincided with the Industrial Revolution. State interests as defined by the metropolitan powers were synonymous with the interests of the industrial plutocrats. Industrialists utilized a third factor of production—capital, essentially, accumulated surpluses—and translated this factor into technological innovation. Mechanization and the harnessing of non-animal sources of energy accelerated production, reduced distances, and thereby condensed time. Within two centuries more change was recorded along the land-people interface than in all prior history combined (Bairoch 1988).

Town and country have become sharply differentiated. Nature could be "visited" during occasional retreats to parks, refuges, zoos, public gardens and reserves. Also, mobility has increased along with the complexity of production and labor demands. Human place was clearly outside of "nature," which was viewed as "wilderness." In the North American experience Modernization occurred when pioneers of European descent moved further through the Western frontier and "civilized," however coarsely, the "Wild West." Wagon trains were followed by passenger railroads powered by coal, the first among fossil fuels to be used extensively in this way. To accommodate these trains tunnels were carved into mountains. Along the way indigenous American tribes were decimated and "pacified." On the Great Plains another species that shared the same habitat with the Plains Indians, buffalo, dwindled perilously close to extinction.

Faster ships were already at sea and the race for even speedier transport became an imperative of international trade. Given the nature of such competition, faster was never fast enough, and society looked to a vertical spatial dimension, air and space, as the new frontier. By that time people of means were removing themselves further from nature through urbanization. The service sector dealt in large part with the human costs of life in cities and the symptoms of "industrial disease." Industrialization and urban growth were mutually entwined owing to numerous economic and other complimentarities. Climate-control and similar innovations were believed to herald the liberation by science of nature-burdened society. Science applied as engineering and technology would successfully transcend the "encumbrances" of nature.

We have learned that this supposed liberation has had a counter-effect, global dysfunction. Who would have thought that the burning of fossil fuels and the release of freon into the atmosphere would have such dangerous effects as ambient pollution and global warming? Who would have thought that the devices for climate control would eventually lead to climates that are out of control?

Who would have thought?

RECOVERING REALITY

There is a contemporary malaise affecting many if not most or all societies, communities and social strata. From Kashmir to Ottawa there is a sense of things gone awry. This contention will, no doubt, be challenged by futurists on the grounds that human nature is human nature and people have always committed barbarities against each other. That may be true, but the far more important question does not concern the savage within us, but rather the assumption that our civilization represents advancement out of barbarity. We have more amenities than ever before, and in many respects modern technology has greatly enhanced life. There has been growth, and in many areas the drudgeries and indignities of an Industrial Age have been assuaged by more liberal approaches to social policy and by technological comforts. On the whole, though, asking whether human civilization has advanced or retreated is a legitimate, perhaps crucial question. Asked another way, can we emerge from global dysfunction to rest assured that the future of our species is secure?

Three other questions related to the above also demand answers, particularly because our world has become so dysfunctional. The dysfunction has forced us to realized that nature will not tolerate planetary abuse born of either the social or environmental limits to growth. The three questions concern the existential dimensions of social and individual life:

1. Do we have a choice concerning the life we want?

2. If so, is the life we are leading the life we want?

3. Which if any option for social change is sustainable in the wake of dysfunction?

For the first time in history we can, indeed we must, confront these issues collectively. We are quite possibly at the brink of creating a wasteland out of the ecosphere. Since the global scale of our agency renders human ignorance a threat to the continued survival of the ecosphere, only concerted and profound social change can heal the planet. In other words, whatever possiblities exist for human "salvation" rely on local action born of global consciousness.

Social change—particularly Modernization—has been a process of blindly groping forward. We are only now realizing the catastrophic results of human agency on this planet over the past two hundred years and increasingly in the past fifty years. Ironically, it is our progress, or anti-progress, that has extended our awareness.

RECOVERY: MULTIPLE TRAJECTORIES

It took the Industrial Revolution, imperialism and modern science to bind the separate trajectories of distinct peoples into a unitary Trajectory of Modernization. I have proposed that the objective of social change is recovery from global dysfunction. What, though are we trying to recover? My response is that we seek a renewed sense of individual and collective self, of wholeness and of continuity. I refer to recovery as *sustainable endogeniety*. The aims of this approach is the recovery of individual and group identity and autonomy, a sense of belonging to a place and its society, and the reintegration of those aspects of our lives that Modernization succeeded in segmenting.

Recovery also means an awareness of painful truths, specifically that we have engaged in a kind of modern alchemy that, in effect, presupposed unlimited growth despite bounded resources and intricately complex biophysical systems. Further, it is possible that the ecosphere has been so damaged by human actions that recovery is no longer possible since our "progress" has reduced biodiversity, landscape integrity and cultural uniqueness.

As discussed in Chapter 1, our capacity for recovery is dependent on reversible adaptations to global dysfunction. Modernization has left us with fewer resources for future choices, and we are poorer for this. The degree of technological mastery we have achieved must rapidly distinguish between sustainable and unsustainable interventions and apply the latter to the recovery of our planet.

Recovery for sustainable endogeneity requires curtailing consumption, particularly extravagant consumption. This in turn will demand life-style changes that will reduce mobility, restore home and workplace as a single unit or proximate units, promote the production of quality, durable goods rather than disposable ones, and renew crafts as a major economic activity. Choices will be limited and privacy compromised. The wealthy societies of the North must redistribute their wealth—internally *and* externally—based on needs-driven criteria and not on the artificial demands of manipulated markets.

It is axiomatic that the number of people consuming resources cannot continue to expand indefinitely. Therefore, curtailing northern appetites for global resources disproportionate to their representation in the world's production is vital to our survivability. Since the skewed distribution of resources derives not from natural endowment but by their transfer from the South to the North, it is incumbent on the latter to take the lead in cutting consumption.

RECOVERING PLACE

Recovery, like dysfunction, derives from global crisis. The road to recovery, though, is not found in that which is uniform. The rediscovery of place and diversity, rather than conformity, is the only way back—or forward—to stability. Even if what we have lost could be termed idyllic, it is a lost paradise unavailable as a future option. Our only recourse is to choose strategies for social change that are sustainable and compatible with the separate futures of autonomous societies and communities.

The individuation of solutions for the future cannot be achieved in isolation. Rather, in a world as interwoven as our own, constant communication and coordination is of fundamental importance. Recovery will depend on smaller-scaled communities and a society that carefully allocates its resources. Simpler technologies appropriate to lower levels of energy expenditure are of the utmost importance.

Also, technical processes such as the production of synthetic materials, aspects of genetic engineering and large-scale transformation of place—including such "empty" places as deserts and tundra—must undergo far greater scrutiny than ever before.

A FULLER VIEW OF THE PARTS OF THE WHOLE

The challenge humanity faces today is as much relational as material. Basic needs must include existential ones including personal fulfillment. Recovery will take great spiritual fortitude, not necessarily in a traditional religious way (though not excluding these paths, either) but in a manner that mobilizes all the will, stamina, intention and faith contained within the individual person and the individual group. Our experience with the development of global dysfunction illuminates the limits to our knowledge and wisdom; recognizing this is not just prudent but imperative. Further, any attempt to suppress human spirit into a single ultra-Modern cast of what human beings "should be" must never be repeated. The violence of imposed conformity is intolerable.

Our ecosphere cannot be atomized into parcels of water, soil and air that are independent of each other. The global commons must be dealt with globally, collectively, between societies and within them. Scarce resources cannot be managed as market commodities since they are easily squandered under that system. These resources sustain life and as such they cannot be subjected to further risk. Just as there is little prospect for preserving biodiversity without a commitment to respecting cultural diversity, no local solution to environmental problems is sufficient to stop global dysfunction.

Reversing local manifestations of global dysfunction is an inherently social enterprise. The only sustainable ecology is a social ecology where relations among people, societies and our environments are conducted with care, equity and foresight. We will have to live lives considerably different from, but potentially much more rewarding, than those available to us today.

To the extent that this endeavor is social, it is political as well. The privileged elites who value their empires above all else will surely defy the necessary steps toward socialization of common resources. Sadly, those who have concluded that there is no other sustainable path available to us will have to resist such defiance with proportional force. Hopefully, moral suasion will be the most efficacious force in this struggle.

The struggle for ecological equity is a matter of survival. The only alternative is the end of the earth and its life-forms. If the latter choice is taken, our bones will be rattling forever in a cold, lifeless abyss with no prospect for recovery or continuity.

That is a burden humanity should not have to bear.

NOTES

1. The Modernization schema offered here does not imply the representation of historical "truth." It is a framework for conceptualizing the interrelated effects of social change and its evolution. It should be considered an epistemic tool only.

Bibliography

Abed, George T., ed. 1988. *The Palestinian Economy: Studies in Development Under Prolonged Occupation*. London and NY: Routledge.

Abdel-Fadil, Mahmoud. 1987. The Macro-Behavior of Oil-Rentier States in the Arab Region. In *The Rentier State, V. II: The Foundations of the Arab State,* edited by Beblawi Hazem and Giacomo Luciani. London, NY: Croom Helm.

Abrahamian, Ervand. 1982. *Iran Between Two Revolutions*. Princeton, N.J.: Princeton University Press.

Adams, Michael, ed. 1988. *Handbooks to the Modern World: The Middle East*. NY: Facts on File.

Adamson, David. 1965. *The Kurdish War*. NY, Washington: Frederick A. Praeger.

Adedeji, Adebayo, ed. 1981. *Indigenization of African Economies*. NY: Africana Publishing.

Agnew, John A. 1989. Sameness and Difference: Hartshorne's The Nature of Geography and Geography as Areal Variation. In *Reflections on Richard Hartshorne's The Nature of Geography,* edited by J. Nicholas Entriki and Stanely D. Brunn. Washington, DC: Association of American Geographers.

———. 1982. Sociologizing the geographical imagination: spatial concepts in the world-system perspective. *Political Geography Quarterly* 1(2).

Ajomo, M.A. 1983. An Appraisal of the Organization of the Petroleum Producing Countries. In *The Politics of Middle Eastern Oil,* edited by J.E. Peterson. Washington, DC: Middle East Institute.

Ake, Claude. 1981. Kenya. In *Indigenization of African Economies,* edited by Adebayo Adedeji. NY: Africana Publishing.

Akhbar Ruz Institute. 1983. *Text of the First Economic, Social and Cultural Development Plan of the Islamic Republic of Iran*. Tehran: Akhbar Ruz Institute.

Ali, Taiser Mohammed Ahmed. 1988. The State and Agricultural Policy: In Quest of Framework for Analysis of Development Strategies. In *Sudan: State, Capital and Transformation,* edited by Tony Barnett and Abbas Abdelkarim. London: Croom Helm.

Alonso, William. 1988. Population and Regional Development. In *Regional Economic Development,* edited by Benjamin Higgins and Donald J. Savoie. Boston: Unwin Hyman.

Althusser, Louis. 1971. *Lenin and Philosophy and Other Essays*. NY and London: Monthly Review Press.

Amin, Samir. 1990. *Maldevelopment: Anatomy of a Global Failure*. Tokyo and London: United Nations University Press and Zed Press.

———. 1981. Senegal. In *Indigenization of African Economies,* edited by Adebayo Adedeji. NY: Africana Publishing.

Amin, Samir et al. 1982. *Dynamics of Global Crisis*. NY: Monthly Review Press.

Amirahmadi, Hooshang. 1987. A theory of ethnic collective movements and its application to Iran. *Ethnic and Racial Studies 10(4).*

Amnesty International. 1982. Campaign against torture—in Turkey: torture is widespread. *Amnesty Action,* July / August .

———. Amnesty International Asks to Visit Turkish Military Prison. *Amnesty International News Release*. NY, August 17.

Amuzegar, Jahangir. 1983. Oil Wealth: A Very Mixed Blessing. In *The Politics of Middle Eastern Oil,* edited by J.E. Peters. Washington, DC: Middle East Institute.

Anderson, Benedict. 1983. *Imagined Communities: Reflections on the Origin and Spread of Nationalism*. London: Verso Editions.

Anderson, James N. 1987. Lands at Risk, People at Risk: Perspectives on Tropical Forest Transformations in the Philippines. In *Lands at Risk in the Third World,* edited by Peter D. Little, Michael M. Horowitz and A. Endre Nyerges. Boulder, CO: Westview Press.

———. 1986a. On theories of nationalism and the size of states. *Antipode* 18(2).

———, ed. 1986b. *The Rise of the Modern State*. Atlantic Highlands, NJ: Humanities Press International, Inc.

Antonious, George. 1979. *The Arab Awakening*. NY: Paragon.

Arfa, Hassan. 1966. *The Kurds: An Historical and Political Study*. London: Oxford University Press.

Arjomand, Said Amir. 1988. *The Turban for the Crown: The Islamic Revolution in Iran*. NY, Oxford: Oxford University Press.

Arnot, R. Page. 1924. *The Politics of Oil: An Example of Imperialist Monopoly*. London: The Labour Publishing Company.

Bachrach, Jere L. 1984. *A Middle East Studies Handbook*. Seattle, WA: University of Washington Press.

Baffoe, Frank. 1981. Southern Africa. In *Indigenization of African Economies,* edited by Adebayo Adedeji. NY: Africana Publishing.

Bairoch, Paul. 1988. *Cities and Economic Development: From the Dawn of History to the Present*. Translated by Christopher Braider. Chicago: University of Chicago Press.

Bakhash, Shaul. 1989. Historical Setting. In *Iran, A Country Study,* edited by Helen Chapin Metz. Washington, DC: Federal Research Division, Library of Congress.

Balandier, Georges. 1970. *Political Anthropology*. Translated from the French by A.M. Sheridan Smith. NY: Pantheon Books.

Balsan, Francois. 1945. *The Sheep and the Chevrolet: A Journey Through Kurdistan*. London: Paul Elek.

Baran, P. 1957. *The Political Economy of Growth*. New York and London: Monthly Review Press.

Barnett, Tony and Abbas Abdelkarim, eds. 1988. *Sudan: State, Capital and Transformation*. London, NY: Croom Helm.

Barnett, Tony. 1989. *Social and Economic Development: An Introduction*. New York and London: The Guilford Press.

Bartelmus, Peter. 1986. *Environment and Developmen*t. Boston: Allen and Unwin.

Barth, F. 1969. *Ethnic Groups and Boundaries: The Social Organizing of Culture and Differences*. Boston: Little and Brown.

———. 1956. "Ecologic Relationships of Ethnic Groups in Swat, North Pakistan." *American Economist* 58:1079–89.

———. 1953. Principles of Social Organization in Southern Kurdistan. Oslo. Universitetets Ethnografiske Museum Bulletin No. 7.

Bartelmus, Peter. 1986. *Environment and Development*. Boston: Allen and Unwin.

Basbananlik, T.C, 1985. *Fifth Five Year Development Plan, 1985-1989.* Ankara: Devlet Planlama Teskilati.

Bassett, Thomas J. 1988. The political ecology of peasant-herder conflicts in the northern Ivory Coast. *Annals of the Association of American Geographers* 78(3):453–472.

Beaumont, Peter, Gerald H. Blake and J. Malcom Wagstaff. 1976. *The Middle East: A Geographic Study*. London, NY: John Wiley and Sons.

Beblawi, Hazem. 1987. The Rentier State in the Arab World. In *The Rentier State, V. II: The Foundations of the Arab State,* edited by Beblawi Hazem and Giacomo Luciani. London, NY: Croom Helm.

Beblawi, Hazem and Giacomo Luciani, eds. 1987. *The Rentier State, V. II: The Foundations of the Arab State.* London: Croom Helm.

Becker, Abraham S., Bert Hansen, and Malcom H. Kerr. 1975. *The Economics and Politics of the Middle East*. NY: American Elsevier Publishing Company, Inc.

Belaid, Sadok. 1988. Role of Religious Institutions in Support of the State. In *Beyond Coercion: The Durability of the Arab State. V. III: The Foundations of the Arab State,* edited by Adeed Dawisha and I. William Zartman . London, NY: Croom Helm.

Bengio, Ofra. 1989. *Mered HaKurdim B'Iraq* (*The Kurdish Rebellion in Iraq,* in Hebrew). Tel Aviv: Kav Adom, Kibbutz Hameucad Publishers for the Dayan Center.

Bennholdt-Thomsen, Veronika. 1982. Subsistence production and extended reproduction. A contribution to the discussion about modes of production. *The Journal of Peasant Studies* 9(4): 241–54.

Bernstein, Richard J. 1976. *The Restructuring of Social and Political Theory.* NY: Harcourt Brace Jovanovich.

Betts, Robert Brenton. 1988. *The Druze.* New Haven and London: Yale University Press.

Beyer, Jacquelyn L. 1980. Africa. In *World Systems of Traditional Resource Management,* edited by Gary A. Klee. London and New York: Edward Arnold and Halsted Press.

Bill, James A. and Carl Leiden. 1974. *The Middle East: Politics and Power.* Boston: Allyn and Bacon, Inc.

Birks, J.S. and C.A. Sinclair. 1983. International Labor Migration in the Gulf. In *The Politics of Middle Eastern Oil,* edited by J.E. Peterson. Washington, DC: Middle East Institute.

Black-Michaud, Jacob, 1986. *Sheep and Land: The Economics of Power in a Tribal Society.* Cambridge: Cambridge University Press.

Blaike, Piers and Harold Brookfield, eds. 1987. *Land Degradation and Society.* London and New York: Methuen.

Blair, John M. 1978. *The Control of Oil.* NY: Vintage Books.

Blake, Gerald H. 1981. Offshore Politics and Resources in the Middle East. In *Change and Development in the Middle East. Essays in Honor of W.B. Fisher,* edited by John I. Clarke and Howard Bowen-Jones. London and NY: Methuen.

Blaut, James M. 1987. *The National Question: Decolonizing the Theory of Nationalism.* London and NJ: Zed Books.

———. 1986. A theory of nationalism. *Antipode* 18(1).

Bonine, Michael Edward. 1980. *Yazd and Its Hinterland: A Central Place System of Dominance in the Central Iranian Plateau*. Marburg/Lahn: Im Selbstverlag des Geographischen Institutes der Universitat Marburg.

Blomstrom, Magnus and Bjorn Hettne. 1984. *Development Theory in Transition: The Dependency Debate and Beyond—Third World Responses*. London: Zed Books.

Bookchin, Murray. 1990. *Remaking Society: Pathways to a Green Future*. Boston: South End Press.

Bois, Thomas. 1966. *The Kurds*. Translated from the French by M.W.M. Wlland. Beirut: Khayats.

Boserup, Ester. 1965. *The Conditions of Agricultural Growth: The Economics of Agrarian Change Under Population Pressure*. Chicago: Aldine.

Brass, Paul, ed. 1985. *Ethnic Groups and the State*. Totowa, NJ: Barnes and Noble.

Brown, James. 1988. The politics of transition in Turkey. *Current History* 87(526).

Brown, Lester R. 1990. The Illusion of Progress. In *State of the World 1990*, edited by Lester R. Brown et al. NY: W.W. Norton.

Brown, Lester R. and Edward C. Wolf. 1985. Soil Erosion: Quiet Crisis in the World Economy. In *Global Ecology*, edited by Charles H. Southwick. Sunderland, MA: Sinauer Associates.

Brown, Lester R., Christopher Flavin, and Hal Kane. 1992. *Vital Signs 1992: The Trends That Are Shaping Our Future*. NY and London: W.W. Norton.

Brush, S.B. and B.L Turner. 1987. The Nature of Farming System. In *Comparative Farming Systems*, edited by B.L. Turner II and S.B. Brush. NY and London: Guilford Press.

Bulloch, John and Harvey Morris. 1992. *No Friends But the Mountains: The Tragic History of the Kurds*. NY and Oxford: Oxford University Press.

Butzer, Karl W. 1989. Cultural Ecology. In *Geography in America*, edited by Gary L. Gaile and Cort J. Willmott. Columbus, OH: Merrill Publishing Co.

Byres. T.J. 1985. Modes of production and non-European pre-colonial societies: the nature and significance of the debate. *The Journal of Peasant Studies* 12(2,3).

Cantori, Louis J. and Iliya Harik, eds. 1984. *Local Politics and Development in the Middle East*. Boulder, CO and London: Westview.

Carlstein, Tommy. 1982. *Time Resources, Society and Ecology: On the Capacity for Human Interaction in Space and Time*. London, Boston: Allen and Unwin.

Carmichael, Joel. 1967. *The Shaping of the Arabs*. NY: Macmillan.

Carpenter, Richard A., ed. 1983. *Natural Systems for Development: What Planners Need to Know*. New York: Macmillan.

Carter, Laraine Newhouse. 1979. Historical Setting [of Iraq]. In *Iraq: A Country Study*, 3rd ed., edited by Richard F. Nyrop. Washington, DC: American University, Foreign Area Studies.

Cernea, Michael M. 1985. *Putting People First*. NY: Oxford University Press for the World Bank.

Chaliand, Gerard. 1980. *People Without A Country, The Kurds and Kurdistan*. Translated by Michael Pallis. London: Zed Press.

Chaliand, Gedrard and Jean-Pierre Rageau. 1985. *Strategic Atlas: A Comparative Geopolitics of the World's Powers*. NY: Harper and Row.

Chambers, Robert. 1983. *Rural Development: Putting the Last First*. London: Longman.

Chileshe, Jonathan H. 1981. Zambia. In *Indigenization of African Economies*, edited by Adebayo Adedeji. NY: Africana Publishing.

Chorley, Richard J. 1973. Geography as Human Ecology. In *Directions in Geography*, edited by Richard J. Chorley. NY and London: Methuen.

Clark, Gordon L., and Michael Dear. 1984. *State Apparatus, Structures and Language of Legitimacy*. Boston: Allen and Unwin Inc.

Clarke, John I., and Howard Bowen-Jones. 1981. *Change and Development in the Middle East. Essays in Honor of W.B. Fisher*. London and NY: Methuen.

Clay, Jason W. 1992. Resource Wars: Nation-State Conflicts of the 20th Century. In *Growing Our Future: Food, Security and the Environment,* edited by Katie Smith and Tetsunao Yamamori. West Hartford, CT: Kumarian Press.

Cobbah, Josiah A.M. 1988. Toward a Geography of Peace in Africa: Redefining Sub-State Self-Determination Rights. In *Nationalism, Self-Determination and Political Geography,* edited by R.J. Johnston, David B. Knight and Eleonore Kofman. London: Croom Helm.

Cohen, Saul B. 1975. *Geography and Politics in a World Divided*. NY: Oxford University Press.

Collelo, Thomas, ed. 1988. *Syria, A Country Study*. Washington, DC: Federal Research Division, Library of Congress.

Commoner, Barry. 1971. *The Closing Circle: Nature, Man and Technology*. NY: Alfred A. Knopf

Costanza, Robert, ed. 1991. *Ecological Economics: The Science and Management of Sustainability*. NY: Columbia University Press.

Cooke, Philip. 1989. Nation, Space, Modernity. In *New Models in Geography, The Political-Economy Perspective,* Volume I, edited by Richard Peet and Nigel Thrift. London: Unwin Hyman.

Cooper, Charles A. and Sidney S. Alexander, eds. 1972. *Economic Development and Population Growth in the Middle East*. NY: American Elsevier.

Copans, Jean. 1980. From Senegambia to Senegal: The Evolution of Peasantries. In *Peasants in Africa: Historical and Contemporary Perspectives,* edited by Martin A. Klein. Beverly Hills, CA: Sage Publications.

Coser, Lewis A. 1968. *Continuities in the Study of Social Conflict*. NY: The Free Press.

Curtis, Michael, ed. *Religion and Politics in the Middle East*. Boulder, CO: Westview.

Davis, Eric. 1983. The Political Economy of the Arab Oil-Producing Nations: Convergence with Western Interests. In *The Politics of Middle Eastern Oil*, edited by J.E. Peterson. Washington, DC: Middle East Institute.

Dawisha, Adeed. 1987. Arab Regimes: Legitimacy and Foreign Policy. In *Beyond Coercion: The Durability of the Arab State. V. III: The Foundations of the Arab State*, edited by Adeed Dawisha and I. William Zartman . London, NY: Croom Helm.

Degefe, Befekadu. 1981. Ethiopia. In *Indigenization of African Economies*, edited by Adebayo Adedeji. NY: Africana Publishing.

Deutsch, Karl W. 1966. *Nationalism and Social Communication: An Inquiry into the Foundations of Nationality*. Cambridge, MA: M.I.T. Press.

De Vos, George and Lola Romanucci-Ross, eds. 1975. *Ethnic Identity: Cultural Continuities and Change*. Palo Alto, CA: Mayfield Publishing Company.

Drysdale, Alasdair. 1988. National Integration Problems in the Arab World: The Case of Syria. In *Nationalism, Self-Determination and Political Geography*, edited by R.J. Johnston, David B. Knight and Eleonore Kofman . London: Croom Helm.

Drysdale, Alasdair and Gerald H. Blake. 1985. *The Middle East and North Africa: A Political Geography*. NY and London: Oxford University Press.

Durning, Alan B. 1989. Ending Poverty. In *State of the World 1990*, edited by Linda Starke. A World Watch Institute Report on Progress Toward a Sustainable Society. NY: W.W. Norton and Co.

Eagleton, William Jr. 1963. *The Kurdish Republic of 1946*. London: Oxford University Press.

Edmonds, C.J. 1957. *Kurds, Turks and Arabs: Politics, Travel and Research in North-Eastern Iraq, 1919-1925*. London: Oxford University Press.

Edokpayi, S.I. 1981. Administrative and Managerial Implications. In *Indigenization of African Economies,* edited by Adebayo Adedeji. NY: Africana Publishing.

Eglin, Darrel R. 1979. The Economy, Government and Politics. In *Iraq: A Country Study,* 3rd ed., edited by Richard F. Nyrop. Washington, DC: American University,

Ehrilch, Paul R., Anne H. Ehrlich and John P. Holdren. 1991. *Healing the Planet: Strategies for Resolving the Environmental Crisis.* Reading, MA: Addison-Wesley.

———. 1973. *Human Ecology: Problems and Solutions.* San Francisco: W.H. Freeman and Company. Eickelman, Dale F. 1981. *The Middle East: An Anthropological Approach.* Englewoods Cliffs, NJ: Prentice-Hall, Inc.

Elazar, Daniel J. 1978. Federalism, Governance, and Development. In *The Third World: Premises of U.S. Policy,* edited by W. Scott Thompson. San Francisco: Institute for Contemporary Studies.

Ellen, Roy. 1982. *Environment, Subsistence and System: The Ecology of Small-Scale Social Formations.* Cambridge: Cambridge University Press.

Ellis, H. and H. Wallich, eds. 1962. *Economic Development for Latin America.* NY: St. Martin's Press.

Emel, Jacque and Richard Peet. 1989. Resource Management and Natural Hazards. In *New Models in Geography, The Political-Economy Perspective,* Volume I, edited by Richard Peet and Nigel Thrift. London: Unwin Hyman.

Emmanuel, A. 1972. *Unequal Exchange.* London and New York: Monthly Review Press.

Enloe, Cynthia H. 1980. *Ethnic Soldiers: State Security in Divided Societies.* Athens, GA: The University of Georgia Press.

———. 1973. *Ethnic Conflict and Political Development.* Boston: Little, Brown and Company.

Entelis, John P. 1981. Ethnic Conflict and the Emergence of Radical Christian Nationalism in Lebanon. In *Religion and Politics in the Middle East,* edited by Michael Curtis. Boulder, CO: Westview.

Entessar, Nader. 1989. The Kurdish mosaic of discord. *Third World Quarterly* 11(4).

Entrikin, J. Nicholas and Stanly D. Brunn, eds. 1989. *Reflections on Richard Hartshorne's The Nature of Geography*. Washington, DC: Association of American Geographers.

Escobar, Arturo. 1992. Planning. In *The Development Dictionary: A Guide to Knowledge as Power*, edited by Wolfgang Sachs. London and NJ: Zed Books.

Esman, Milton J. and Itamar Rabinovich, eds. 1988. *Ethnicity, Pluralism, and the State in the Middle East*. Ithaca and London: Cornell University Press in Cooperation with the Dayan Center for Middle Eastern and African Studies at Tel Aviv University.

Esteva, Gustavo. 1992. Development. In *The Development Dictionary: A Guide to Knowledge as Power*, edited by Wolfgang Sachs. London and NJ: Zed Books.

Europa Publications Limited. 1989. *The Middle East and North Africa 1989*. London: Europa Publications Limited.

———. 1986. *The Middle East and North Africa 1986*. London: Europa Publications Limited.

Ezeife, Emeka. 1981. Nigeria. In *Indigenization of African Economies*, edited by Adebayo Adedeji. NY: Africana Publishing

Farouk-Sluglett, Marion, Peter Sluglett and Joe Stork. 1984. Not Quite Armageddon: Impact of the War on Iraq. MERIP Reports. No. 125/126, Vol. 14; No. 6/7, July–September.

Feili, Omram Yahya and Arlene R. Fromchuck. 1976. The Kurdish struggle for independence. *Middle East Review* 5(9).

Feshbech, Murry. 1993. Environmental Calamities: Widespread and Costly. In The Former Soviet Union in Transition, edited by Richard Kaufman and John Hardt. Armonic, NY and London: M.E. Sharpe.

Fink, Daniel B. 1993. *Judaism and Ecology*. NY: Hadassah.

Fisher, W.B. 1988a. Egypt. *The Middle East and North Africa*. London: Europa Publications.

———. 1988b. Turkey. *The Middle East and North Africa*. London: Europa Publications.

———. 1988c. Jordan. *The Middle East and North Africa*. London: Europa Publications.

———. 1988d. Lebanon. *The Middle East and North Africa*. London: Europa Publications.

———. 1978. *The Middle East, A Physical, Social and Regional Geography*, 7th ed. London: Methuen and Co.

Forde, C. Daryll. 1934. *Habitat, Economy and Society: A Geographical Introduction to Ethnology*. NY: E.P. Dutton and Co., Inc.

Fowler, Catherine S. 1977. Ethnoecology. In *Ecological Anthropology*, edited by Donald L. Hardesty. NY: John Wiley and Sons.

Frank, Andre Gundar. 1981. *Crisis In the Third World*. NY: Holmes and Meier.

———. 1966. The development of underdevelopment. *Monthly Review*.

Friedmann, John. 1992. *Empowerment: The Politics of Alternative Development*. Cambridge, MA and Oxford, UK: Blackwell.

———. 1966. *Regional Development Policy: A Case Study of Venezuela*. Cambridge, MA. and London: The M.I.T. Press.

Friedmann, John and Clyde Weaver. 1979. *Territory and Function, The Evolution of Regional Planning*. London: Edward Arnold.

Friedmann, John and Mike Douglass. 1981. Agropolitan Development: Towards a New Strategy for Regional Planning in Asia. In *Growth Pole Strategy and Regional Development Policy*, edited by F. Lo and Kamal Salih, 163–192. Oxford: Pergamon Press.

Friedmann, John and Yvone Forest. 1988. The Politics of Place: Toward A Political Economy of Territorial Planning. In *Regional Economic Development*, edited by Benjamin Higgins and Donald J. Savoie. Boston: Unwin Hyman.

Gabbay, Shoshana, ed. 1994. *The Environment in Israel*. Jerusalem: Ministry of the Environment.

Gage, Nicolas. 1979. Iranian Kurds Return to Own Struggle. *The New York Times,* March 1, 1979, p. A3.

Geertz, Clifford. 1963. *Agricultural Involution: the Process of Ecological Change in Indonesia.* Berkeley, CA: University of California Press.

Gellner, Ernest 1983. The Tribal Society and Its Enemies. In *The Conflict of Tribe and State in Iran and Afghanistan,* edited by Richard Tapper. NY: St. Martin's Press.

Gesellschaft fur Bedrohte Volker. 1987. Minorities in the Gulf War. *Cultural Survival Quarterly* 11(4).

Ghanem, Shukri. 1986. *OPEC: The Rise and Fall of an Exclusive Club.* London, NY and Sydney: KPI, distributed by Routledge and Kegan Paul.

Ghassemlou, Abdul Rahman. 1980. Kurdistan in Iran. In *People Without A Country, The Kurds and Kurdistan,* edited by Gerard Chaliand. Translated by Michael Pallis. London: Zed Press.

———. 1965. *Kurdistan and the Kurds.* London: Collet's Publishers Ltd.

Glaeser, Bernhard, ed. 1987. *The Green Revolution Revisited: Critiques and Alternatives.* London: Allen and Unwin.

Glazer, Nathan and Daniel P. Moynihan, eds. 1975. *Ethnicity: Theory and Experience.* Cambridge, MA: Harvard University Press.

Gordon, David C. 1971. *Self-Determination and History in the Third World.* Princeton, NJ: Princeton University Press.

Gotlieb, Yosef. 1996. Irreconcilable Planning: The Transformation of Life-Place into Economic Space. *Progress in Planning.* Forthcoming.

———. 1995. *Dysfuncion Mundial.* Fundacion UNA. San Jose, Costa Rica: National University of Costa Rica. Forthcoming.

———. 1992. The influence of ethnic segmentation on development. *Geography Research Forum* 12: 81–101.

———. 1991. The Geo-Ethnic Imperatives of Development: The Inter-Dynamics of Territory, Society and State in the Third World.

Worcester, MA: Graduate School of Geography, Clark University. Unpublished dissertation.

———. 1982. *Self-Determination in the Middle East.* NY: Praeger.

———. 1981. Sectarianism and the Iraqi State. In *Religion and Politics in the Middle East,* edited by Michael Curtis. Boulder, CO: Westview.

Graham, Robert. 1978. *Iran: The Illusion of Power.* NY: St. Martin's Press.

Green, Reginal-Herbol. Foreign direct investment and African political economy. 1981. In *Indigenization of African Economies,* edited by Adebayo Adedeji. NY: Africana Publishing.

Grillo, Ralph and Alan Rew. 1985. *Social Anthropology and Development Policy.* London and NY: Tavistock Publications.

Grimaldi, Fulvio. 1983. Dawn Three: A Mountain Raid—The Middle East. T*he Middle East* 108 (October): 24–25.

Gunter, Michael M. 1988. The Kurdish problem in Turkey. *The Middle East Journal.* 42(3).

Hanafi, Mohammed N. Egypt. 1981. In *Indigenization of African Economies,* edited by Adebayo Adedeji. NY: Africana Publishing.

Hardesty, Donald L. 1977. *Ecological Anthropology.* NY: John Wiley and Sons.

Harik, Iliya. 1987. The Origins of the Arab State System. In *Nation, State and Integration in the Arab World, V. I: The Foundations of the Arab State,* edited by Ghassan Salame. London, NY: Croom Helm.

Harriss, John. 1982. *Rural Development.* London: Hutchingson University Library.

Hartunian, Vartan. 1968. *Neither to Laugh Nor to Weep.* Boston: Beacon Press.

Harthshorne, Richard. 1939. The nature of geography: a critical survey of current thought in light of the past. *Annals of the Association of American Geographers* 29: 3–4.

Harvey, David. 1990. Between space and time: reflections on the geographical imagination. *Annals of the Association of American Geographers* 80: 3.

Hawley, Amos H. 1986. *Human Ecology: A Theoretical Essay*. Chicago: The University of Chicago Press.

Held. Colbert C. 1989. *Middle East Patterns: Places, Peoples, and Politics*. Boulder: Westview Press.

Helms, Christine Moss. 1984. *Iraq: Eastern Flank of the Arab World*. Washington, DC: The Brookings Institution.

Hempstone, Smith. 1979. Khomeini's "Feast of Blood." *The New Republic*, September 22, 1979.

Hermassi, Elbaki. 1987. State-Building and Regime Performance in the Greater Maghreb. In *Nation, State and Integration in the Arab World, V. I: The Foundations of the Arab State*, edited by Ghassan Salame. London, NY: Croom Helm.

Higgins, Benjamin and Donald J. Savoie, eds. 1988. *Regional Economic Development*. Boston: Unwin Hyman.

Hilhorst, Jos G.M. 1971. *Regional Planning: A Systems Approach*. Rotterdam: Rotterdam University Press.

Hirschmann, A.O. 1981. *Trespassing: Economics to Politics and Beyond*. Cambridge: Cambridge University Press.

———. 1958. *The Strategy of Economic Development*. New Haven, CT: Yale University Press.

Hooglund, Eric. 1989. In *Iran, A Country Study*, edited by Helen Chapin Metz. Washington, DC: Federal Research Division, Library of Congress.

Horowitz, Donald L. 1985. *Ethnic Groups in Conflict*. Berkeley and Los Angeles: University of California Press.

Howe, Marvine. 1981. Turks Imprison Former Minister Who Spoke up on Kurds' Behalf. *The New York Times*. March 27, 1981.

———. 1980. Turkish Military Warns Politicians to Unite In Solving Country's Ills. *The New York Times*. January 3, 1980, p. A1.

Hughes, Arnold. 1981. The Nation-State in Black Africa. In *The Nation-State: The Formation of Modern Politics*, edited by Leonard Tivey. NY: St. Martin's Press.

Human Rights Internet. 1981. The repression of the Kurds in Turkey. *HRI Reporter*, 6 (May-June).

Hunter, Guy, A.H. Bunting and Anthony Bottrall, eds. 1980. Policy and Practice in Rural Development: Proceedings of the Second International Seminar on Change in Agriculture. London: Croom Helm for Overseas Development Council.

Hunter, Shireen, ed. 1985. *Political and Economic Trends in the Middle East: Implications for U.S. Policy*. Boulder, CO: Westview.

Ibrahim, Ibrahim, ed. 1983. *Arab Resources: The Transformation of a Society*. Washington, DC: Center for Contemporary Arab Studies; London: Croom Helm.

Iltis, Hugh H. 1985. Tropical Forests: What Will Be Their Fate? In *Global Ecology*, edited by Charles H. Southwick. Sunderland, MA: Sinauer Associates.

Independent Commission on International Development Issues. 1986 ,1980. *North-South: A Program for Survival*. Cambridge, MA: The MIT Press.

International Commission of Jurists. 1980. The trial of Mihri Belli and the TEP Party. *The Review of The International Commission of Jurists*. 24 (June).

International League for Human Rights. 1977. Iraqi Persecution of Kurds. *Annual Report, 1976–1977*.

Isbell, Billie Jean. 1985. *To Defend Ourselves: Ecology and Ritual in an Andean Village*. Prospect Heights, IL: Waveland Press.

Islamoglu-Inan, Huri, ed. 1987. The Ottoman Empire and the World-Economy. Cambridge, NY: Cambridge University Press.

Islamoglu-Inan, Huri and Caglar Keydar. 1987. Agenda for Ottoman History. In *The Ottoman Empire and the World-Economy*, edited by Huri Islamoglu-Inan. Cambridge, NY: Cambridge University Press.

Ismael, Jacqueline. 1983. The Economic Transformation of Kuwait. In *The Politics of Middle Eastern Oil,* edited by J.E. Peterson. Washington, DC: Middle East Institute.

Issawi, Charles. 1982. *An Economic History of the Middle East and North Africa.* NY: Columbia University Press.

Izady, Mehrdad R. 1992. *The Kurds: A Concise Handbook.* Washington, DC: Taylor and Francis.

Jafar, Majeed R. 1976. Under-Underdevelopment: A Regional Case Study of the Kurdish Area in Turkey, No. 24. Helsinki: Studies of the Social Policy Association in Finland.

Jawad, Sa'ad. 1982. Recent Developments in the Kurdish Issue. In *Iraq: The Contemporary State,* edited by Tim Niblock. NY: St. Martin's Press.

Jochim, Michael A. 1981. *Strategies for Survival: Cultural Behavior in an Ecological Context.* NY: Academic Press.

Johnson, Douglas L. 1969. The Nature of Nomadism: A Comparative Study of Pastoral Migrations in Southwestern Asia and Northern Africa. Research Paper No. 118. Department of Geography, The University of Chicago.

Johnson, E.A.J. 1970. *The Organization of Space in Developing Countries.* Cambridge, M.A.: Harvard University Press.

Johnston, R.J. 1989. *Environmental Problems: Nature, Economy and the State.* London and NY: Belhaven Press.

———. 1989. The State. Political Geography, and Geography. In *New Models in Geography, The Political-Economy Perspective,* Volume I, edited by Richard Peet and Nigel Thrift. London: Unwin Hyman.

Johnston, R.J., David B. Knight and Eleonore Kofman, eds. 1988. *Nationalism, Self-Determination and Political Geography.* London: Croom Helm.

Johnston, R.J. and P.J. Taylor, eds. 1986. *A World in Crisis? Geographical Perspectives.* Oxford, U.K.: Basil Blackwell.

Kamenka, Eugene, ed. 1983. *The Portable Karl Marx.* NY: Viking Penguin.

———, ed. 1976. *Nationalism: The Nature and Evolution of an Idea*. NY: St. Martin's Press.

Kasaba, Resat. 1988. *The Ottoman Empire and the World Economy: The Nineteenth Century*. Albany, NY: State University of New York.

Kashi, Ed. 1994. *When the Borders Bleed: The Struggle of the Kurds*. NY: Pantheon.

Kasperson, Roger E. and Julian V. Minghi, eds. 1969. *The Structure of Political Geography*. Chicago: Aldine.

Kazemi, Farhad. 1988. Ethnicity and the Iranian Peasantry, eds. 1988. In *Ethnicity, Pluralism, and the State in the Middle East*, edited by Milton J. Esman and Itamar Rabinovich. Ithaca and London: Cornell University Press in Cooperation with the Dayan Center for Middle Eastern and African Studies at Tel Aviv University.

Kechichian, Joseph A. 1990. National Security. In *Iran, A Country Study*, edited by Helen Chapin Metz. Washington, DC: Federal Research Division, Library of Congress.

Keddie, Nikki R. with a section by Yann Richard. 1981. *Roots of Revolution. An Interpretative History of Modern Iran*. New Haven and London: Yale University Press.

Kedourie, Elie, ed. 1970. *Nationalism in Asia and Africa*. NY: New American Library.

Kelidar, A.R. 1983. The Problem of Succession in Saudi Arabia. In *The Politics of Middle Eastern Oil*, edited by J.E. Peters. Washington, DC: Middle East Institute.

Kelkar, Govind and Dev Nathan. 1991. *Gender and Tribe: Women, Land and Forests*. London, UK and Atlantic Highlands, NJ: Zed Books.

Kendal. 1981. Kurdistan in Turkey. In *People Without A Country, The Kurds and Kurdistan*, edited by Gerard Chaliand. Translated by Michael Pallis. London: Zed Press.

———. 1980. The Kurds Under the Ottoman Empire. In *People Without A Country, The Kurds and Kurdistan*, edited by Gerard Chaliand. Translated by Michael Pallis. London: Zed Press.

Kendeigh, S. Charles. 1975. Man's Biological Environment. In *Human Ecology,* edited by Norman D. Levine. North Scituate, MA: Duxbury Press.

Keyder, Caglar. 1987. *State and Class in Turkey: A Study in Capitalist Development*. London, NY: Verso.

Keys, Charles F. 1981. *Ethnic Change*. Seattle, WA: University of Washington Press.

Khalidi, Rashid. 1988. Social Transformation and Political Power in the Radical Arab States. In *Beyond Coercion: The Durability of the Arab State. V. III: The Foundations of the Arab State,* edited by Adeed Dawisha and I. William Zartman. London, NY: Croom Helm.

Khotari, Rajni. 1989. Forward to *Staying Alive: Women, Ecology and Development,* by V. Shiva. London and NJ: Zed Books.

Kimche, Jon. 1975. Selling out the Kurds. *New Republic,* V (172): 19.

———. 1970. *The Second Arab Awakening*. NY: Holt, Rinehart, and Winston.

Kinnane, Derk. 1964. *The Kurds and Kurdistan*. London, NY: Oxford University Press.

Kirby, Andrew. 1985. Pseudo-random thoughts on space, scale and ideology in political geography. *Political Geography Quarterly* 4(1).

Klee, Gary A., ed. 1980. *World Systems of Traditional Resource Management*. London and New York: Edward Arnold and Halsted Press.

Knight, David B. 1988. Self-Determination for Indigenous Peoples: The Context for Change. In *Nationalism, Self-Determination and Political Geography,* edited by R.J. Johnston, David B. Knight and Eleanore Kofman. London: Croom Helm.

———. 1982. Identity and territory: geographical perspectives on nationalism and regionalism. *Annals of the Association of American Geographers*. 72(4).

Knight, Gregory C. 1974. *Ecology and Change: Rural Modernization in an African Community*. NY, London: Academic Press.

Knox, Paul and John Agnew. 1989. *The Geography of the World Economy*. London and New York: Edward Arnold.

Knuttson, Karl Eric. 1969. Dichotomization and Integration. In *Ethnic Groups and Boundaries: The Social Organizing of Culture and Differences,* edited by F. Barth. Boston: Little and Brown.

Kuznets, Simon. 1965. *Economic Growth and Structure.* London: Heinemann.

Kuklinski, Antoni R. 1975. *Regional Development and Planning: International Perspectives.* Amsterdam: Sijthoff-Leydent.

———, ed. 1972. *Growth Poles and Growth Centers in Regional Planning.* Paris, The Hague: Mouton.

Kurian, George Thomas. 1982. *The Encyclopedia of the Third World.* NY: Facts on File.

Kutschera, Chris. 1983. A Jail Away from Home. *The Middle East* No. 99 (January): 33.

Lanne, Peter. 1977. *Armenia. The First Genocide of the 20th Century.* Munich: Institute for Armenian Studies.

Lea, David A.M. and D.P. Chaudhri, eds. 1983. *Rural Development and the State.* London and New York: Methuen.

Leach, E.R. 1940. *Social and Economic Organization of the Rowandux Kurds.* London: Percy Lund, Humphries and Co. Ltd. for The London School of Economics.

Lenczowski, George. 1975. *Political Elites in the Middle East.* Washington: DC: American Enterprise Institute for Public Policy Research.

———. 1952. *The Middle East in World Affairs.* Ithaca, NY: Cornell University Press.

Lenin, V.I. 1915. The Imperialist War: The Struggle Against Social-Chauvinism and Social-Pacificism. In *The Collected Worlds of V.I. Lenin.* NY: International Publishers.

Leonard, H. Jeffrey. 1989. Overview: Environment and the Poor—Development Strategies for a Common Agenda. In *Environment and the Poor: Development Strategies for A Common Agenda,* edited by Leonard. H. Jeffrey et al. Overseas Development Council,

U.S.-Third World Policy Perspectives, No. 11. New Brunswick, NJ and Oxford: Transaction Books.

Levine, Norman D., ed. 1975. *Human Ecology*. North Scituate, MA: Duxbury Press.

Lewis, Mark. 1990. Historical Setting. In *Iran, A Country Study*, edited by Helen Chapin Metz. Washington, DC: Federal Research Division, Library of Congress.

Lewis, W.A. 1954. Economic Development with Unlimited Supply of Labor. Paper, The Manchester School, May.

Lipset, Seymour Martin. 1978. Racial and Ethnic Tensions in the Third World. In *The Third World: Premises of U.S. Policy*, edited by W. Scott Thompson. San Francisco: Institute for Contemporary Studies.

Little, Ian M.D. 1982. *Economic Development: Theory, Policy, and International Relations*. NY: Basic Books.

Lo, F.C. and K. Salih. 1978. *Growth Pole Strategy and Regional Development Policy*. Oxford: Pergamon Press

Longrigg, Stephen. 1963. *The Middle East, A Social Biography*. Chicago: Aldine.

Longino, Helen E. 1990. *Science as Social Knowledge. Values and Objectivity in Scientific Inquiry*. Princeton, NJ: Princeton University Press.

Lopez, Maria Elena. 1987. The Politics of Lands at Risk in a Philippine Frontier. In *Lands at Risk in the Third World*, edited by Peter D. Little, Michael M. Horowitz and A. Endre Nyerges. Boulder, CO: Westview Press.

MacDonald, Charles G. 1988. The Kurdish Question in the 1980s. In *Ethnicity, Pluralism, and the State in the Middle East*, edited by Milton J. Esman and Itamar Rabinovich. Ithaca and London: Cornell University Press in Cooperation with the Dayan Center for Middle Eastern and African Studies at Tel Aviv University.

MacLaughlin, Jim. 1986. State-centered social science and the anarchist critique: ideology in political geography. *Antipode* 18(1).

MacNeill, Jim. 1990. Strategies for Sustainable Economic Development. In *Managing Planet Earth,* edited by Scientific American. NY: W.H. Freeman.

Markusen, Ann R. 1987. *Regions: The Economics and Politics of Territory.* Totowa, NJ: Rowman and Littlefield.

MacPherson, Angus. 1989. The Economy. In *Iran, A Country Study,* edited by Helen Chapin Metz. Washington, DC: Federal Research Division, Library of Congress.

Malek, Mohammed H. 1989. Kurdistan in the Middle East conflict. *New Left Review.* 17 (May/June 1989).

Marr, Phebe. 1985. *The Modern History of Iraq.* Boulder, CO: Westview Press.

Marx, Karl. 1845–46. The German Ideology, Volume 1. As excerpted in *The Portable Karl Marx,* edited by Eugene Kamenka. NY: Viking Penguin.

———. 1844. On the Jewish Question. In *The Portable Karl Marx,* edited by Eugene Kamenka. NY: Viking Penguin.

Marx, Karl and Fredrich Engels. 1967. *The Communist Manifesto.* NY: Washington Square Press.

Mason, Robert Scott. 1990. The Economy. In *Iran, A Country Study,* edited by Helen Chapin Metz. Washington, DC: Federal Research Division, Library of Congress.

———. 1988. The Society and Its Environment. In *Syria, A Country Study,* edited by Thomas Collelo. Washington, DC: Federal Research Division, Library of Congress.

Max-Neef, Manfred A. et al. 1991. *Human-Scale Development: Conception, Application and Further Reflections.* NY and London: The Apex Press.

Mazrui, Ali A. 1972. *Cultural Engineering and Nation-Building in East Africa.* Evanston, IL: Northwestern University Press.

McDowall, David. 1992. *The Kurds. A Nation Denied.* London: Minority Rights Group.

———. 1989. *The Kurds.* Report No. 23. London: Minority Rights Group.

McGowan, Afaf Sabeh. 1988. Historical Setting. In *Syria, A Country Study,* edited by Thomas Collelo. Washington, DC: Federal Research Division, Library of Congress.

McKee, David L. et al. 1970. *Regional Economics: Theory and Practice.* NY: The Free Press.

McKibben, Bill. 1989. *The End of Nature.* NY: Random House.

Mennes, Tinbergen, Waardenburg. 1979. *Spatial Approach to Modernization, The Element of Space in Development Planning.* The Hague: Mouten.

Mercer, David. 1987. Patterns of protest: native lands rights and claims in Australia. *Political Geography Quarterly* 6(2).

Merchant. Carolyn. 1992. *Radical Ecology: The Search for a Livable World.* NY, London: Routledge.

———. 1989. *Ecological Revolutions: Nature, Gender, and Science in New England.* Chapel Hill and London: The University of North Carolina Press.

Messerschmidt, Donald A. 1987. Conservation and Society in Nepal. In *Lands at Risk in the Third World: Local-Level Perspectives,* edited by Peter D. Little and Michael M. Horowitz et al. Boulder, CO: Westview.

Metz, Helen Chapin, ed. 1990, 1989. *Iraq, A Country Study.* Washington, DC, Federal Research Division, Library of Congress.

Minority Rights Group. 1989. *Massacre By Gas.* London: A Minority Rights Group Profile.

Morris, M.D. 1976. *A Physical Quality of Life Index.* NY: Praeger for the Overseas Development Council.

Moore, Barrington, Jr. 1966. *Social Origins of Dictatorship and Democracy: Lord and Peasant in the Making of the Modern World.* Boston: Beacon.

Muhammad, Nesrin. 1980. Caught in the middle for the nth time. *The Middle East*. No. 73 (November): 18–19.

Murdoch, William W. 1980. *The Poverty of Nations: The Political Economy of Hunger and Population*. Baltimore and London: The Johns Hopkins University Press.

Murra, John. 1976. *Formaciones Economicas y Politicas Del Mundo Andino*. Lima: Institudo de Estudios Peruvanos.

Murray, Andrew. 1975. The Kurdish struggle. *Patterns of Prejudice* 9(July/August).

Murton, Brian J. 1980. South Asia. In *World Systems of Traditional Resource Management*, edited by Gary A. Klee. London and New York: Edward Arnold and Halsted Press.

Mushi, S.S. 1981. Kenya. In *Indigenization of African Economies*, edited by Adebayo Adedeji. NY: Africana Publishing.

Myrdal, Gunnar. 1968. *Asian Drama: An Inquiry into the Poverty of Nations*. NY: The Twentieth Century Fund.

———. 1960. *Beyond the Welfare State: Economic Planning and Its Economic Implications*. New Haven: Yale University Press.

———. 1957. *Rich Lands and Poor: The Road to World Prosperity*. NY: Harper and Brothers.

Naamani, Israel T. 1968. The Kurds in Iraq. *Jewish Frontier*, July-August.

Najmabadi, Afsaneh, 1987. Depoliticization of a Rentier State: The Case of Pahlevi Iran. In *The Rentier State, V. II: The Foundations of the Arab State*, edited by Beblawi Hazem and Giacomo Luciani. London, NY: Croom Helm.

Nawawi, Mohammed A. 1969. Stagnation as a basis of regionalism: A lesson from Indonesia. *Asian Survey* V, IX (12).

Nazdar, Mustafa. 1980. The Kurds in Syria. In *People Without A Country, The Kurds and Kurdistan*, edited by Gerard Chaliand. Translated by Michael Pallis. London: Zed Press.

Nelan, Bruce W. 1992. A land of stones. *Time*, March 2: 28.

Netting, Robert. 1981. *Balancing on an Alp: Ecological Change and Continuity in a Swiss Mountain Community.* New York: Cambridge University Press.

New Kurdish coalition. *The Middle East.* No. 96 (October) :11.

New York Times. 1977. Iraq Accused of Trying to Wipe Out Kurds. *The New York Times,* January 15, 1977.

Niblock, Tim, ed. 1982. *Iraq: The Contemporary State.* NY: St. Martin's Press.

Nonneman, Gerd. 1988. *Development Administration and Aid in the Middle East.* London and NY: Routledge.

Nyrop, Richard F., ed. 1979. *Iraq, A Country Study,* 3rd ed. Washington, DC: American University, Foreign Area Studies.

O'Ballance, Edgar. 1973. *The Kurdish Revolt: 1961–1970.* Hamden, CT: Archon Press.

Odell, Peter R. 1983. The Significance of Oil. In *The Politics of Middle Eastern Oil,* edited by J.E. Peterson. Washington, DC: Middle East Institute.

Odum, H.T. 1981. *Systems Ecology.* NY: J. Wiley and Sons.

———. 1971. *Fundamentals of Ecology,* 3rd ed. Philadelphia: W.B. Saunders Company.

O'Laughlin, John. 1986. World-Power Competition and Local Conflicts in the Third World. In *A World in Crisis? Geographical Perspectives,* edited by Johnston, R.J. and P.J. Taylor. Oxford: Basil Blackwell Ltd.

Olson, Robert. 1989. *The Emergence of Kurdish Nationalism and the Sheikh Said Rebellion, 1880-1925.* Austin, TX: University of Texas Press.

Ophuls, William and A. Stephen Boyan, Jr. 1992. *Ecology and the Politics of Scarcity Revisited: The Unraveling of the American Dream.* NY: W.H. Freeman.

O'Riordan, Timothy. 1989. The Challenge for Environmentalism. In *New Models in Geography, The Political-Economy Perspective,*

Volume I, edited by Richard Peet and Nigel Thrift. London: Unwin Hyman.

Owusu-Ansah, K.A. 1981. Ghana. In *Indigenization of African Economies*, edited by Adebayo Adedeji. NY: Africana Publishing.

Pamuk, Sevket. 1989. Ottoman Agriculture, 1840-1913. In *The Ottoman Empire and the World-Economy*, edited by Huri Islamoglu-Inan. Cambridge, NY: Cambridge University Press.

Parker, Geoffrey. 1985. *Western Geopolitical Thought in the Twentieth Century*. NY: St. Martin's Press.

Parkes, Don and Nigel Thrift. 1980. *Times, Spaces and Places: A Chronographic Perspective*. Chinchester: John Wiley and Sons.

Patai, Raphael. 1969. *Society, Culture and Change in the Middle East*, 3rd ed. Philadelphia: University of Pennsylvania Press.

Peet, Richard. 1986. The Destruction of Regional Cultures. In *A World in Crisis? Geographical Perspectives*, edited by R.J. Johnston and P.J. Taylor. Oxford: Basil Blackwell.

Peet, Richard and Nigel Trift, eds. 1989. *New Models in Geography*. London: Unwin Hyman.

Pelletiere, Stephen. 1990. The Society and Its Environment. In *Iran, A Country Study*, edited by Helen Chapin Metz. Washington, DC: Federal Research Division, Library of Congress.

Penrose, Edith. International Oil Companies and Governments in the Middle East. In *The Politics of Middle Eastern Oil*, edited by J.E. Peterson. Washington, DC: Middle East Institute.

Pitman, Paul M. III, ed. 1988. *Turkey, A Country Study*. Washington, DC: Federal Research Division, Library of Congress.

Plan and Budget Organization. 1984. *A Statistical Reflection of the Islamic Republic of Iran*. Tehran: Statistical Center of Iran.

Plaut, W. Gunther, Bernard J. Bamberger and William W. Hallo. 1981. *The Torah: A Modern Commentary*. NY: Union of American Hebrew Congregation.

Portugali, Juval. 1988. Nationalism, Social Theory and the Israel/ Palestinian Case. In *Nationalism, Self-Determination and Political Geography*, edited by R.J. Johnston, et al. London: Croom Helm.

Popper, Karl R. 1983. *Realism and the Aim of Science*. Totowa, NJ: Rowman and Littlefield.

———. 1959. *The Logic of Scientific Discovery*. London: Hutchinson.

Posey, Darrell Addison. 1985. Indigenous management of tropical forest ecosystems: the case of the Kayapo Indians of the Brazilian Amazon. *Agroforestry Systems* 3:139–58.

Preston, Natheniel Stone. 1967. *Politics, Economics and Power: Ideology and Practice Under Capitalism, Socialism, Communism, and Fascism*. NY: Macmillan.

Pye, Lucian W., ed. 1963. *Communications and Political Development*. Princeton, NJ: Princeton University Press.

Raha'i Translation Group. 1981. Kurdistan: Resistance and Further Prospects. Number 2, Spring-Summer. Rahai, Tehran: Organization of Communist Unity.

Rappaport, Roy A. 1984. *Pigs for Ancestors: Ritual in the Ecology of a New Guinea People*, 2nd ed. New Haven, CT: Yale University Press.

Redclift, Michael. 1987. *Sustainable Development: Exploring the Contradictions*. London and NY: Routledge.

Rees, Judith. 1985. *Natural Resources: Allocation, Economics and Policy*, 2nd ed. London and New York: Routledge.

Rejwan, Nessim. 1980. The Kurds: Khomeini's hidden time bomb. *Hadassah*, April.

Revolutionary Communist Party, USA. 1982. Film and Revolution: A Talk with Yilmaz Guney. *Revolutionary Worker*, October 1, page 1.

Rhodes, Robert I. 1970. *Imperialism and Underdevelopment: A Reader*. NY: Monthly Review Press.

Richards, Alan, ed. 1986. *Food, States and Peasants: Analysis of the Agrarian Question in the Middle East*. Boulder, CO and London: Westview Press.

Richards, Paul. 1985. *Indigenous Agricultural Revolution: Ecology and Food Production in West Africa.* Boulder, CO: Westview Press.

Richardson, Harry W. 1973. *Regional Growth Theory.* London: Macmillan .

Riddell, Robert. 1981. *Ecodevelopment.* Westmead, Farnborough, Hampshire, UK: Gower Publishing.

Robinson, Walter V. 1991. Kurds Flee as Rebellion Crumbles. *The Boston Globe,* April 4, 1991, p. 1.

Rocheleau, Dianne E . 1991. Gender, ecology and the science of survival: stories and lessons from Kenya. *Agriculture and Human Values* 8 (1).

———. 1987. Women, Trees and Tenure: Implications for Agroforestry Research and Development. In *Land, Trees and Tenure,* edited by J.B. Raintree. Nairobi and Madison, WI: ICRAF/LTC.

Ronen, Dov. 1979. *The Quest for Self-Determination.* New Haven, CT: Yale University Press.

Roosevelt, Jr., Archie. 1980. The Kurdish Republic of Mahabad. In *People Without A Country, The Kurds and Kurdistan,* edited by Gerard Chaliand. Translated by Michael Pallis. London: Zed Press.

Roseberry, William. 1988. Political economy. *Annual Review of Anthropology* 17:161–85.

Rostow, W.W. 1960. *The Stages of Economic Growth.* Cambridge, UK: Cambridge University Press.

Roszak, Theodore. 1992. *The Voice of the Earth: An Exploration of Ecopsychology.* NY: Touchstone.

Ruckelshaus, William D. 1990. Toward a Sustainable World. In *Scientific American,* 1990. *Managing Planet Earth.* NY: W.H. Freeman.

Sachs, Ignacy. 1987. Towards a Second Green Revolution. In *The Green Revolution Revisited: Critiques and Alternatives,* edited by Bernhard Glaeser. London: Allen and Unwin.

Sachs, Wolfgang et al., eds. 1992. *The Development Dictionary: A Guide to Knowledge as Power.* London and NJ: Zed Books.

Sack, Robert David. 1987. Human Territoriality and Space. Wallace W. Atwood Lecture Series No. 3, The Graduate School of Geography. Worcester, MA: Clark University.

Salame, Ghassan. 1987. "Strong" and "Weak" States, a Qualified Return to the Muqaddimah. In *Nation, State and Integration in the Arab World, V. I: The Foundations of the Arab State,* edited by Ghassan Salame. London, NY: Croom Helm.

Saeedpour, Vera Beaudin and Ismet Scherif Vanly. 1981. *The Kurds.* Cultural Survival Inc. 5/1.

Safrastian, Arshak. 1948. *Kurds and Kurdistan.* London: The Harvill Press.

Sauer, Carl O. 1981. *Selected Essays 1963–1975.* Berkeley, CA: Turtle Island Foundation for the Netzahaulcoyotl Historical Society.

———. 1956. The Agency of Man on Earth. In *Man's Role in Changing the Face of the Earth,* edited by W.L. Thomas. Chicago: University of Chicago Press.

Sayer, Andrew. 1984. *Method in Social Science: A Realist Approach.* London: Hutchinson.

Sayigh, Yusif A. 1982. *The Arab Economy: Past Performance and Future Prospects.* Oxford: Oxford University Press.

Sbert, José Maria. 1992. Progress. In *The Development Dictionary: A Guide to Knowledge as Power,* edited by Wolfgang Sachs et al. London and NJ: Zed Books.

Schmidt, Dana Adams. 1964. *Journey Among Brave Men.* Boston: An Atlantic Monthly Press Book.

Scientific American. 1990. *Managing Planet Earth.* NY: W.H. Freeman.

Secretariat of the Economic Commission for Asia and the Far East. 1984. Towards Integration in Asia. In *Economic Integration and Third World Development, edited* by Paradip K. Ghosh. Westport, CT: Greenwood Press.

Seddon, David. 1989. Economy (of Turkey). In *The Middle East and North Africa 1989,* 35th ed. London: Europa Publications.

Seidman, Ann. 1986. *Money, Banking and Public Finance in Africa.* London and NJ: Zed Books Ltd.

Seidman, Robert B. 1978. *The State, Law and Development.* NY: St. Martin's Press.

Seers, Dudley. 1983. *The Political Economy of Nationalism.* Oxford: Oxford University Press.

Sharabi, Hisham. 1983. The Poor Rich Arabs. In *Arab Resources: The Transformation of a Society,* edited by Ibrahim Ibrahim. Washington, DC: Center for Contemporary Arab Studies; London: Croom Helm.

Shibutani, Tamotsu and Kian M. Kwan. 1965. *Ethnic Stratification: A Comparative Approach.* NY: Macmillan.

Shiva, Vandana. 1991. *The Violence of the Green Revolution: Third World Agriculture, Ecology and Politics. London* and NJ: Zed Books Ltd.

———. 1989. *Staying Alive: Women, Ecology and Development.* London and NJ: Zed Books Ltd.

Shiva, Vandana et al. *Biodiversity: Social and Ecological Perspectives.* London and NJ: Zed Books and Penang, Malaysia: World Rainforest Movement.

Short, John R. 1982. *An Introduction to Political Geography.* London: Routledge and Kegan Paul.

Sim, Richard. 1980. Kurdistan: The Search For Recognition. Conflict Studies. No. 124, November.

Skocpol, Theda. 1979. *States and Social Revolutions: A Comparative Analysis of France, Russia, and China.* Cambridge: Cambridge University Press.

Slater, David. 1989. Territorial power and the peripheral state: the issue of decentralization. *Development and Change* 20: 501–31.

Smith, Anthony D. 1988. *The Ethnic Origins of Nations.* NY: Basil Blackwell.

———. 1972. *Theories of Nationalism.* NY: Harper Torchbooks.

Smith, Neil. 1984. *Uneven Development: Nature, Capital and the Production of Space.* Oxford, UK and Cambridge, MA: Basil Blackwell.

Snyder, Louis L. 1982. *Global Mini-Nationalisms: Autonomy or Independence.* Westport, CT: Greenwood Press.

———, ed. 1964. *The Dynamics of Nationalism.* NY: D. Van Nostrand Company.

Soiffer, Stephen M. and Gary N. Howe. 1982. Patrons, clients and the articulation of modes of production: an examination of the penetration of capitalism into the peripheral agriculture in northeastern Brazil. *The Journal of Peasant Studies* 9(2).

Spooner, Brian. 1987. Insiders and Outsiders in Baluchistan: Western and Indigenous Perspectives on Ecology and Development. In *Lands at Risk in the Third World,* edited by Peter D. Little, Michael M. Horowitz and A. Endre Nyerges. Boulder, CO: Westview Press.

Stea, David and Ben Wisner. 1984. The fourth world, a geography of indigenous struggles. *Antipode* 16(2).

Steinhart, Carol E. 1975. Antecedents and Forces of Socio-Cultural Evolution. In *Human Ecology,* edited by Norman D. Levine. North Scituate, MA: Duxbury Press.

Stephens, Robert. 1976. *The Arab's New Frontier.* Boulder, CO: Westview Press.

Steward, Julian H. 1976. *Theory of Culture Change: The Methodology of Multilinear Evolution,* 3rd ed. Urbana, Chicago: University of Illinois Press.

———, ed. 1967. *Contemporary Change in Traditional Societies.* Urbana, Chicago: University of Illinois Press.

Stöhr, Walter B. and D.R. Fraser Taylor, eds. 1981. *Development from Above or Below? The Dialectics of Regional Planning in Developing Countries.* Chinchester and NY: John Wiley and Sons.

Stork, Joe. 1982. State Power and Economic Structure. In *Iraq: The Contemporary State,* edited by Tim Niblock. NY: St. Martin's Press.

Stavrianos, L.S. 1981. *Global Rift: The Third World Comes of Age*. NY: William Morrow.

Streeten, Paul, et al. 1981. *First Things First: Meeting Basic Human Needs in Developing Countries.* NY: Oxford University Press for the World Bank.

Sunar, Ilkay. 1987. State and Economy in the Ottoman Empire. In *The Ottoman Empire and the World-Economy,* edited by Huri Islamoglu-Inan. Cambridge, NY: Cambridge University Press.

Swartz, Marc J., Victor W. Turner and Arthur Tuden, eds. 1966. *Political Anthropology*. Chicago: Aldine Publishing Company.

Szentes, Tzmas. 1973. *The Political Economy of Underdevelopment.* Budapest: Akademiai Kiado.

Tanzer, Michael. 1974. *The Energy Crisis: World Struggle for Power and Wealth.* NY and London: Monthly Review Press.

Tapper, Richard, ed. 1983. *The Conflict of Tribe and State in Iran and Afghanistan*. NY: St. Martin's Press.

Tartter, Jean R. 1988. National Security. In *Turkey, A Country Study,* edited by Paul M Pitman, III. Washington, DC, Federal Research Division, Library of Congress.

Taylor, Charles Lewis and Michael C. Hudson, 1972. *World Handbook of Political and Social Indicators,* 2nd ed. New Haven, CT: Yale University Press.

Taylor, Charles Lewis and David A. Jodice. 1983. World Handbook of Political and Social Indicators, 3rd ed. New Haven, CT: Yale University Press.

Taylor, Paul W. 1986. *Respect For Nature: A Theory of Environmental Ethics.* Princeton, NJ: Princeton University Press.

Taylor, Peter J. 1986. An exploration into world-systems—analysis of political parties. *Political Geography Quarterly*. Supplement to 5(4).

Taylor, Peter and John House, eds. 1984. *Political Geography: Recent Advances and Future Directions*. Totowa, NJ: Barnes and Noble Books.

Telberg, V.G. 1965. *Telberg's Translation to Atlas Narodov Mira*. NY: Telberg Book Corporation.

Thomas, Clive Y. 1974. *Dependence and Transformation: The Economics of the Transition to Socialism*. NY and London: Monthly Review Press.

Thompson, W. Scott, eds. 1978. *The Third World: Premises of U.S. Policy*. San Francisco: Institute for Contemporary Studies.

Tinker, Hugh. 1981. The Nation-State in Asia. In *The Nation- State: The Formation of Modern Politics,* edited by Leonard Tivey. NY: St. Martin's Press.

Tivey, Leonard, ed. 1981. *The Nation-State: The Formation of Modern Politics*. NY: St. Martin's Press.

Toffler, Alvin, 1989 (1980). *The Third Wave*. NY: Bantam Books.

Tucker, William F. 1989. Introduction to *The Emergence of Kurdish Nationalism and the Sheikh Said Rebellion, 1880-1925,* by Olson, Robert. Austin, Texas: University of Texas Press.

Turner, B.L. II. 1989. The specialist-synthesis approach to the revival of geography: the case of cultural ecology. *Annals of the Association of American Geographers* 79(1): 88–100.

Turner, B.L. II and S.B. Brush, eds. 1987. *Comparative Farming Systems*. NY: Guilford Press.

Ul Haq, Mahbub. 1981. Preface in *First Things First: Meeting Basic Human Needs in Developing Countries, by* Paul Streeten et al., eds. NY: Oxford University Press for the World Bank.

———. 1976. *The Poverty Curtain: Choices for the Third World*. NY: Colombia University Press.

UNESCO. 1988. *1988 Statistical Yearbook*. Paris: United Nations Educational, Scientific and Cultural Organization.

UPI. 1990. Iraqis Are Said to Execute Kurds. *The Boston Globe*, World Briefs, Thursday, June 21, 1990.

van Bruinessen, M.M. 1984. The Kurds in Turkey. *MERIP Reports*. No. 121. Vol. 14, No. 2. February.

———. 1983. Kurdish Tribes and the State of Iran: The Case of Simko's Revolt. In *The Conflict of Tribe and State in Iran and Afghanistan*, edited by Richard Tapper. NY: St. Martin's Press.

———. 1978. Agha, Shaikh and State: On the Social and Political Organization of Kurdistan. A Doctoral Dissertation. Utrecht: Rijksuniversiteit Te Utrecht.

Vanly, Ismett Cheriff. 1980. Kurdistan in Iraq, 1980. In *People Without A Country, The Kurds and Kurdistan*, edited by Gerard Chaliand. Translated by Michael Pallis. London: Zed Press.

———. 1968. *The Kurdish Problem in Syria: Plans for the Genocide of a National Minority*. n.p. Committee for the Defence of the Kurdish People's Rights.

Verhelst, Thierry G. 1990. No *Life Without Roots: Culture and Development*. London and NJ: Zed Books.

Voices from exile. *The Middle East*. 92 (June): 27.

Wallerstein, I. 1982. "Crisis as Transition." In *Dynamics of Global Crisis*, edited by S. Amin. NY and London: Monthly Review Press.

Ward, Barbara. 1973. Only One World. In *Who Speaks for Earth*, edited by M. Strong. NY: W.W. Norton and Company.

Warriner, Doreen. 1948. *Land and Poverty in the Middle East*. London and NY: Royal Institute of International Affairs.

Watts, M.J. 1983a. On the Poverty of Theory: Natural Hazard Research in Context. In *Interpretations of Calamity*, edited by K. Hewitt. London, Sydney: Allen and Unwin.

———. 1983b. *Silent Violence: Food, Famine and Peasantry in Northern Nigeria*. Berkeley, CA.: University of California Press.

Weaver, Clyde. 1981. Development Theory and the Regional Question: A Critique of Spatial Planning and its Detractors. In *Development from Above or Below? The Dialectics of Regional Planning in Developing Countries*, edited by Walter B. Stöhr and D.R.F. Taylor. Chinchester and New York: John Wiley and Sons.

Weitz, Raanan. 1986. *New Roads to Development: A Twentieth Century Fund Essay*. NY: Greenwood Press.

———. 1971. *From Peasant to Farmer: A Twentieth Century Fund Study.* NY: Columbia University Press.

Whitaker, Ben. 1973. *The Fourth World.* NY: Schocken.

Whittlesey, D. 1969. The Regional Concept and the Regional Method. In *American Geography: Inventory and Prospect*, edited by James and Jones.

Wilson, Edward O. 1990. Threats to Biodiversity (in *Scientific American). Managing Planet Earth.* NY: W.H. Freeman.

Wilson, Rodney. 1979. *The Economies of the Middle East.* NY: Holmes and Meier Publishers, Inc.

Wisner, Ben. 1988. *Power and Need in Africa.* London: Earthscan Publications.

Wittfogel, Karl A. 1929. Translated by G.L. Ulmen. *Antipode*.17 (1).

Woodson, Jr., LeRoy. 1975. We who face death. *National Geographic* 147 (March).

World Bank. 1994. *World Development Report 1994.* Washington, DC: International Bank for Development and Reconstruction.

———. 1993. *World Development Report 1993.* Washington, DC: International Bank for Development and Reconstruction.

———. 1992. *World Development Report 1992.* Washington, DC: International Bank for Development and Reconstruction.

———. 1991. *World Development Report 1991.* Washington, DC: International Bank for Development and Reconstruction.

———. 1990. *World Development Report 1990.* Washington, DC: International Bank for Development and Reconstruction.

———. 1989. *World Development Report 1989.* Washington, DC: International Bank for Development and Reconstruction.

———. 1988. *World Development Report 1988.* Washington, DC: International Bank for Development and Reconstruction.

———. 1975. *The Assault on World Poverty: Problems of Rural Development, Education and Health.* Baltimore and London: The Johns Hopkins University Press for the World Bank.

World Commission on Environment and Development. 1987. *Our Common Future.* Oxford and NY: Oxford University Press.

World Marxist Review Symposium. The nationalities question in Asian and African countries. *World Marxist Review* 26(3).

World Resources Institute. 1992. *World Resources 1992–93.* A report issued in collaboration with The United Nations Environmental Programme and The United Nations Development Programme. NY and Oxford: Oxford University Press.

Wriggins, Howard. 1966. National Integration. In *Modernization: The Dynamics of Growth,* edited by Myron Weiner. NY and London: Basic Books.

Xenos, Nicholas. 1989. *Scarcity and Modernity.* London and New York: Routledge.

Yapp, M.E. 1987. *The Making of the Modern Middle East, 1792-1923.* London and NY: Longman.

Young, John. 1990. *Sustaining the Earth.* Cambridge, MA: Harvard University Press.

Index

A

Adaptation 4, 19, 21, 34, 45, 48, 49, 50, 54, 74, 82, 84, 89, 93, 94, 143
Africa 66, 105
Agricultural involution 65
Agriculture 72, 76, 93, 138
Agropolitan development 75
Amazonia 67
Anthropocentrism 8
Anthropology 43, 47, 48, 49, 51, 54, 65
Atomism 70
Autonomy 4, 19, 37, 38, 55, 60, 77, 78, 106, 108, 125, 126, 128, 131, 132, 143

B

Bookchin 24, 54
  Social Ecology 53. *See also* Regional political ecology.
Bottom-up regional development 73
Bretton Woods Conference 22

C

Capitalism 9, 11, 15
Categorization 5
Central place theory 86
Climate 11, 15, 84, 125, 141
Colonized Space 92
Community 4, 9, 48, 63, 75, 81, 88, 133
Conservation 67, 75, 81
Consumption 3, 9, 10, 11, 23, 83, 99, 112, 144
Continuity 4, 33, 35, 63, 66, 87, 92, 96, 100, 125, 143, 146
Core-periphery 4, 127
Crises 10, 15, 53, 103, 132, 134
Cultural ecology 43, 48, 49, 50, 54, 65
Culture and nature 45

D

Dead Sea 86
Decolonization 35, 39, 96, 97
Deforestation 15, 29, 57, 137
De-Kurdification 85, 116, 128
Democracy 21, 77
Development 1–5, 8, 9, 11–13, 15, 19–22, 39, 69, 71–82, 84, 86, 89–94, 97, 98, 103–105, 116, 118, 124, 127, 128
  contemporary 22
  growth-oriented 8, 39, 79
  industrialization 2, 8, 10, 15, 29, 39, 71, 72, 75, 76, 116, 141
  irreversible processes 15

objective 9
progress 8, 9, 13, 18, 44, 64, 70, 94, 136, 143
regional 86
secularization 2, 8
socioeconomic advance 8, 97
state-centered 8
Third World 4, 26, 28, 32, 35, 39, 40, 75, 91, 92
urbanization 2, 8, 10, 39, 72, 75, 76, 141
Disempowerment 1
Distribution 77, 90, 99
development funds 124
of energy 46
of Kurds 110
of resources 11, 47, 86, 144
population 118

E

Earth's carrying capacity 65
Eco-development 20, 54, 55, 79
Ecofeminism 54
Ecological
anthropology 43, 47, 48, 51, 54, 65
literacy, concept of 22
Ecological destruction 39, 56
Ecology 9, 11, 19, 20, 22, 24, 40, 43, 45, 46, 48–54, 58–60, 65, 66, 80, 82, 84–87, 93, 94, 98, 104, 105, 118, 125, 145. *See also* ecology, various types; human, social, political.
Economics 10, 36, 82
conventional 27
development 36
ecological 20
growth 10, 15, 40
market 11
neo-classical 70
positivist 27
Ecoregional approach 73, 80, 125
Ecosphere 8, 15, 18, 22, 42, 61, 137, 142, 143, 145
Ecosystem 15, 18, 20, 21, 33, 45, 46, 48, 49, 50, 51, 54, 61, 64, 80, 91, 92, 105, 136
approach 49–51
Education 72, 75, 124
Endogenous change 4
Endogenous recovery 21, 22, 40, 62, 64, 69, 73, 78, 82–84, 85, 87, 92, 93–95, 97, 98, 103–105, 129, 131, 132
Energetics 65
Energy production 11, 99
Environment 89, 90, 91, 93, 94, 98, 99, 103, 104, 135, 138, 140, 145
Environmental
crises 10, 53
degradation 1, 2, 25, 28, 29, 32, 91, 97, 102, 103, 125, 135
determinism 44, 45, 86
determinists 89, 98
limits to growth 142
stresses 138
sustainability 132
Ethnic
conflict 32, 35, 102, 104
groups 106
segmentation 106
Ethnically divided societies 106
Ethnicity, definition of 42
Ethnoecology 59
Ethno-nationalism 88, 100–102, 108
Ethno-political violence 1
Ethnoscience 54, 55, 57–59, 60–62, 90 103
Evolution of development theory 3
Exploitation 1, 67, 74

F

Failure
  of development 1, 12, 42
  of positivist economics 27
Feminization of poverty 1
Fourth World 55, 69, 88
Fractionalization 106
Freedom
  of choice 53
  of cultural expression 78
  of individual 9
Fuel 24, 28, 30, 31, 70, 116, 137, 138, 141

G

Gaia theory 54
Gaza Strip 86
Gender 60
  patterns 28, 84, 92, 93
Geography 33, 48, 63
  ethno 59
  human 43, 90
  political 37
Global
  change 13, 19
  dysfunction 12, 13, 15–22, 25, 26, 34, 39–41, 43, 62, 64, 131, 133, 136, 141–143, 145. *See also* Sustainable endogenous responses.
  problematic 2, 12
Goldenweiser, Alexander 65
Green Revolution 23
Growth 11, 13, 23, 46, 74, 140, 142
  and development 8–9
  and progress 9
  economic 8–11, 15, 25, 27, 36, 71, 73, 78, 82, 86, 97, 118
  employment 25
  limits to 9, 11, 23, 47, 135, 142
  market economies 135
  models of 19
  poles 71, 75, 76, 79, 86
    approach 86
  population 11, 26, 29, 46, 47, 67
  regional 51
  theories of 23
  unlimited 143
  urban 141

H

Habitat 11, 33, 45, 46, 49, 56, 61, 62, 78, 80, 89, 91, 108, 126, 138, 141
Health 28, 29, 72, 99, 123
Heritage 4, 18, 42, 54, 55
High Yielding Varieties 23
Holism 50, 73
Human
  control over nature 44
  ecology 9, 24, 43, 45, 46, 52, 54, 59, 66
  reproduction 61
Humanity 2, 8, 18, 20, 47, 50, 53, 134, 135, 142, 146
Hypergrowth 13, 16, 39, 42, 70, 136

I

Improvement 1, 23, 25
Indigenous knowledge 34, 54, 55, 57, 58, 60–62, 66, 85, 90, 99, 103, 105, 125
  belief systems 66
  land management 67
Industrialization. *See* Development: industrialization.
Iran 107, 108, 111, 118, 121, 122, 123, 124, 125, 128, 129, 132
Iranian Kurdistan 121, 132

Iraq 85, 107, 108, 111, 112, 118, 119, 120, 121, 125, 128, 129, 132
Iraqi Kurdistan 120, 121
Irreversibility 19, 24
Irreversible processes 15

J

Jazireh Strip 132
Jews 57, 66, 105
  ecological perspectives 66

K

Kurdish
  autonomy 132
  consciousness 108
  displacement of peasants 85
  economic system 126
  education 124
  ethnonationalism 108, 126
  Front 126
  health care 123
  identity 107–108, 112, 126
  indigenous displacement 85
  labor force 116, 124
  language 125, 129
  lasses 127
  life place 131
  literacy rates 123
  loans and investments 124
  migration 120
  patterns of oppression 128
  peasants in Iraqi Kurdistan 120
  political culture 126
  population 111
  regions 112
  regions in Iran 123
  religious sects 125
  resettlement of Arabs 85
  rurality 120
  social and economic indicators 124
  social ecology 125
  society 112
  tribes 125, 126
  underdevelopment 127
  unity 108
  Worker's Party 112
Kurdistan 22, 84, 85, 88, 92, 107–113, 116, 118–121, 123, 129 131, 132
  "Arabized" 85
  as an endogenous recovery region 107
Kurds 22, 33, 55, 66, 85, 96 106–108, 110–112, 114, 116, 118, 120–129, 131, 132

L

Land 13, 15, 20, 29, 34, 43, 44, 45, 47, 50–52, 54–57, 64, 65, 67, 72, 74, 77, 82, 116, 118, 126–129, 132, 137, 138, 140
Life-place
  concept 23
  displacement 13, 62, 136
Limits 9, 83
  of capitalism 11
  of carrying capacity 10
  of consciousness 138
  of growth 9, 11, 47, 135
    environmental 142
    social 135
  of resources 47
  to knowledge 145
Loss of
  community 133
  self 133
  wisdom 133

M

Malthus, Thomas 47
Marx, Karl 11, 36, 50, 51–53, 56, 57, 65
Marxism 11, 50
Migration 1, 75, 120
Modernization 2–4, 8, 12, 13, 22, 23, 34–42, 69, 70, 72, 75, 78, 82, 83, 85, 112, 116, 131, 133, 136, 138, 140, 143, 146
  heritage of community and continuity 4
  paradigm 22
  "rational" space economies 3
  rejection 4
Multi-national states 106

N

Nation groups 106
Nationalism 15, 32, 70, 88, 92, 100, 101, 102, 108, 126
Natural resources 8, 15
  equitable distribution 86
Needs 3, 9, 11, 20, 73–75, 77, 78, 116, 118, 132, 138, 144, 145
NGOs 72
Non-state nations 106

O

Oil 116, 121, 135, 138
Orr, D.W. 22

P

Patriarchal tyranny 1
Persecuted minorities 106
Political ecology 20, 40, 51
  school 66
Political struggle 104
Politics 36, 54, 70, 77, 78, 79, 104, 105
Pollution 11, 24, 28, 141
Post-colonial states 1, 4, 21, 22, 23, 33, 35, 70, 107, 127, 128
  socio-spatial composition 23
Poverty 1, 2, 25–29, 32, 39
Power centers 4
Production vs. reproduction 67
Progress. *See* Development: progress.

R

Recovering place 144
Recovery 9, 10, 15, 19–22, 27, 40, 42, 69, 70, 73, 78, 80–85, 131, 132, 143, 144, 146
Regional political ecology 40, 51
Resources 8, 9, 10, 11, 15, 18–20, 24, 27, 29, 34, 135, 143–146
  distribution of 11, 144
  extraction 9, 11, 39
Reversibility 19
Rostow, W.W. 23
Rural people's knowledge 66
Rural-urban outmigration 1

S

Scale 84
Secularization. *See* Development: secularization.
Self-determination 19, 21, 33, 35, 39, 69, 78, 85, 88, 93, 97
Social
  crises 132
  ecology 11, 19, 20, 22, 24, 53, 86, 87, 98, 104, 145
  learning 86

Society-
  environment relationship 91
  land unity 56
  nature relationship 43
    materialistic perspective 66
Socioeconomic development 1
  categorization of societies 5
Socio-spatial realities of territory 34
Southeast Anatolia Project 85, 116, 131
Space economies 21
Spatial
  disequilibrium 70
  parameters 83
State 35
State-centered government. *See also* Development: state-centered.
State-centered modernization 38
Statism 70
Strategies 144
Submerged nationalities 106
Sustainability 12, 19, 28, 62, 85, 132, 143–144, 146
Sustainable
  agriculture 34
  development 5, 94
  ecology 145
  endogeneity 21, 22, 144
  endogenous recovery 62, 78, 82
  endogenous responses 20
  population 65
  social change 93
Symbiotic approaches to society-environment relations 45
Symbiotic human-land relations 44
Syria 85, 107, 108, 111, 118, 125, 128, 129, 132
Syrian Kurdistan 118

T

Technology and liberation from nature 140
Territorial development 20, 74, 75, 81
The Third Wave 64
Third World. *See* Development: Third World.
Turkey 85, 108, 111, 112, 114, 115, 116, 125, 128, 129, 132
Turkish
  Kurdistan 118
  state 112

U

Underdevelopment 1, 13, 127, 128
Urbanization. *See* Development: urbanization.

V

Violence 105

W

Welfare 8, 27, 53, 58
World Bank 22